Structural Steelwork:
Design to Limit State Theory

Third edition

Dennis Lam
School of Civil Engineering
The University of Leeds, Leeds, UK

Thien-Cheong Ang
School of Civil and Environmental Engineering
Nanyang Technological University, Singapore

Sing-Ping Chiew
School of Civil and Environmental Engineering
Nanyang Technological University, Singapore

AMSTERDAM • BOSTON • HEIDELBERG • LONDON • NEW YORK • OXFORD
PARIS • SAN DEIGO • SAN FRANCISCO • SINGAPORE • SYDNEY • TOKYO

ELSEVIER
BUTTERWORTH
HEINEMANN

Elsevier Butterworth-Heinemann
Linacre House, Jordan Hill, Oxford OX2 8DP
200 Wheeler Road, Burlington, MA 01803

First published 1987
Reprinted 1988 (with corrections), 1990, 1991
Second edition 1992
Reprinted 1993 (twice), 1994, 1995, 1997, 1998, 1999, 2001, 2002
Third edition 2004

Permissions may be sought directly from Elsevier's Science and
Technology Rights
Department in Oxford, UK: phone: (+44) (0) 1865 843830;
fax: (+44) (0) 1865 853333; e-mail: permissions@elsevier.co.uk.
You may also complete your request on-line via the Elsevier
homepage (http://www.elsevier.com), by selecting 'Customer
Support' and then 'Obtaining Permissions'.

Cover Image: Swiss Re Building – 30 St Mary Axe – The Gherkin.
Photo with kind permission from Grant Smith Photographer

British Library Cataloguing in Publication Data
Lam, Dennis
   Structural steelwork : design to limit state theory. – 3rd ed.
   1. Steel, Structural 2. Building, Iron and steel
   I. Title II. Ang, Paul III. Chiew, Sing-Ping
   624.1'821

   ISBN   0 7506 59122

Library of Congress Cataloguing in Publication Data
A catalogue record for this book is available from the Library of Congress

ISBN   0 7506 59122

For information on all Butterworth-Heinemann publications
visit our website at www.bh.com

Typeset by Newgen Imaging Systems (P) Ltd, Chennai, India
Printed and bound in Great Britain

Contents

Preface to the third edition

This is the third edition of the Structural Steelwork: Design to Limit State Theory by T.J. MacGinley and T.C. Ang, which proved to be very popular with both students and practising engineers. The change of authorship was forced upon by the deceased of Mr T.J. MacGinley. All the chapters have been updated and rearranged to comply with the latest revision of the BS 5950-1:2000 Structural use of steelwork in building - Part 1: Code of practice for design rolled and welded sections, it may be used as a stand-alone text or in conjunction with BS 5950. The book contains detailed explanation of the principles underlying steel design and is intended for students reading for civil and/or structural engineering degrees in universities. It should be useful to final year students involved in design projects and also sufficiently practical for practising engineers and architects who require an introduction to the latest revision of BS 5950. Every topic is illustrated with fully worked examples and problems are also provided for practice.

D.L.

Preface to the second edition

The book has been updated to comply with

BS 5950: Part I: 1990
Structural Use of Steelwork in Building
Code of Practice for Design in Simple and Continuous Construction:
Hot Rolled Sections

A new chapter on portal design has been added to round out its contents. This type of structure is in constant demand for warehouses, factories and for many other purposes and is the most common single-storey building in use. The inclusion of this material introduces the reader to elastic and plastic rigid frame design, member stability problems and design of moment-transmitting joints.

T.J.M.
T.C.A.

Preface to the first edition

The purpose of this book is to show basic steel design to the new limit state code BS 5950. It has been written primarily for undergraduates who will now start learning steel design to the new code, and will also be of use to recent graduates and designers wishing to update their knowledge.

The book covers design of elements and joints in steel construction to the simple design method; its scheme is the same as that used in the previous book by the principal author, *Structural Steelwork Calculations and Detailing*, Butterworths, 1973. Design theory with some of the background to the code procedures is given and separate elements and a complete building frame are designed to show the use of the code.

The application of microcomputers in the design process is discussed and the listings for some programs are given. Recommendations for detailing are included with a mention of computer-aided drafting (CAD).

T.J.M.
T.C.A.

1

Introduction

1.1 Steel structures

Steel frame buildings consist of a skeletal framework which carries all the loads to which the building is subjected. The sections through three common types of buildings are shown in Figure 1.1. These are:

(1) single-storey lattice roof building;
(2) single-storey portal frame building;
(3) medium-rise braced multi-storey building.

These three types cover many of the uses of steel frame buildings such as factories, warehouses, offices, flats, schools, etc. A design for the lattice roof building (Figure 1.1(a)) is given and the design of the elements for the braced multi-storey building (Figure 1.1(c)) is also included. Design of portal frame is described separately in Chapter 9.

The building frame is made up of separate elements—the beams, columns, trusses and bracing—listed beside each section in Figure 1.1. These must be joined together and the building attached to the foundations. Elements are discussed more fully in Section 1.2.

Buildings are three dimensional and only the sectional frame has been shown in Figure 1.1. These frames must be propped and braced laterally so that they remain in position and carry the loads without buckling out of the plane of the section. Structural framing plans are shown in Figures 1.2 and 1.3 for the building types illustrated in Figures 1.1(a) and 1.1(c).

Various methods for analysis and design have been developed over the years. In Figure 1.1, the single-storey structure in (a) and the multi-storey building in (c) are designed by the simple design method, while the portal frame in (b) is designed by the continuous design method. All design is based on the newly revised limit state design code BS 5950-1: 2000: Part 1. Design theories are discussed briefly in Section 1.4 and design methods are set out in detail in Chapter 2.

1.2 Structural elements

As mentioned above, steel buildings are composed of distinct elements:

(1) Beams and girders—members carrying lateral loads in bending and shear;
(2) Ties—members carrying axial loads in tension;

(a) Single-storey lattice roof building with crane

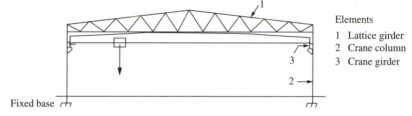

Elements

1   Lattice girder
2   Crane column
3   Crane girder

Fixed base

(b) Single-storey rigid pinned base portal

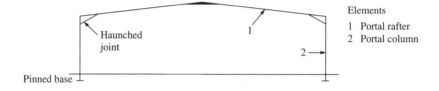

Elements

1   Portal rafter
2   Portal column

Haunched joint

Pinned base

(c) Multi-storey building

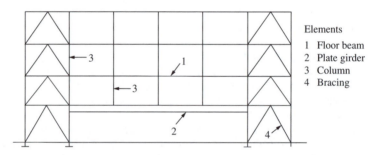

Elements

1   Floor beam
2   Plate girder
3   Column
4   Bracing

Figure 1.1 Three common types of steel buildings

(3) Struts, columns or stanchions—members carrying axial loads in compression. These members are often subjected to bending as well as
compression;
(4) Trusses and lattice girders—framed members carrying lateral loads. These
are composed of struts and ties;
(5) Purlins—beam members carrying roof sheeting;
(6) Sheeting rails—beam members supporting wall cladding;
(7) Bracing—diagonal struts and ties that, with columns and roof trusses, form
vertical and horizontal trusses to resist wind loads and hence provided the
stability of the building.

Joints connect members together such as the joints in trusses, joints between
floor beams and columns or other floor beams. Bases transmit the loads from
the columns to the foundations.

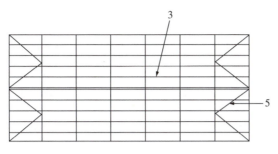

Roof plan

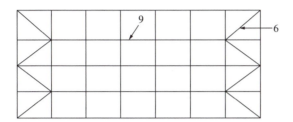

Lower chord bracing

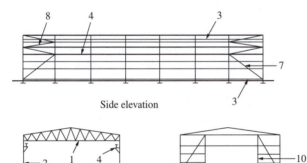

Side elevation

Section   Gable framing

Building elements

1 Lattice girder	6 Lower chord bracing
2 Column	7 Wall bracing
3 Purlins and sheeting rails	8 Eaves tie
4 Crane girder	9 Ties
5 Roof bracing	10 Gable column

Figure 1.2 Factory building

The structural elements are listed in Figures 1.1–1.3, and the types of members making up the various elements are discussed in Chapter 3. Some details for a factory and a multi-storey building are shown in Figure 1.4.

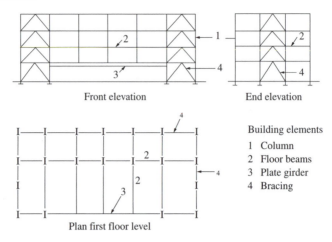

Front elevation   End elevation

Plan first floor level

Building elements
1 Column
2 Floor beams
3 Plate girder
4 Bracing

Figure 1.3 Multi-storey office building

1.3 Structural design

Building design nowadays usually carried out by a multi-discipline design team. An architect draws up plans for a building to meet the client's requirements. The structural engineer examines various alternative framing arrangements and may carry out preliminary designs to determine which is the most economical. This is termed the 'conceptual design stage'. For a given framing arrangement, the problem in structural design consists of:

(1) estimation of loading;
(2) analysis of main frames, trusses or lattice girders, floor systems, bracing and connections to determine axial loads, shears and moments at critical points in all members;
(3) design of the elements and connections using design data from step (2);
(4) production of arrangement and detail drawings from the designer's sketches.

This book covers the design of elements first. Then, to show various elements in their true context in a building, the design for the basic single-storey structure with lattice roof shown in Figure 1.2 is given.

1.4 Design methods

Steel design may be based on three design theories:

(1) elastic design;
(2) plastic design;
(3) limit state design.

Elastic design is the traditional method and is still commonly used in the United States. Steel is almost perfectly elastic up to the yield point and elastic theory is a very good method on which the method is based. Structures are analysed by elastic theory and sections are sized so that the permissible stresses are not

(a) Factory building

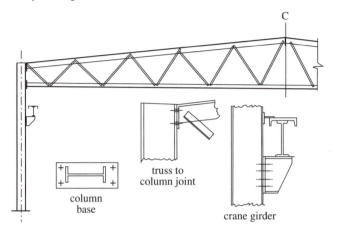

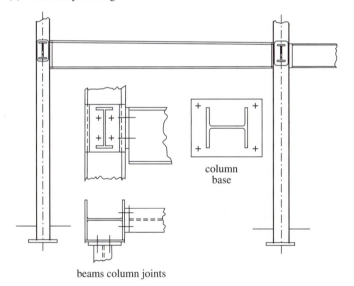

Figure 1.4 Factory and multi-storey building

exceeded. Design in accordance with BS 449-2: 1967: *The Use of Structural Steel in Building* is still acceptable in the United Kingdom.

Plastic theory developed to take account of behaviour past the yield point is based on finding the load that causes the structure to collapse. Then the working load is the collapse load divided by a load factor. This too is permitted under BS 449.

Finally, limit state design has been developed to take account of all conditions that can make the structure become unfit for use. The design is based on the actual behaviour of materials and structures in use and is in accordance with BS 5950: *The Structural Use of Steelwork in Building;* Part 1—Code of Practice for Design—Rolled and Welded Sections.

The code requirements relevant to the worked problems are noted and discussed. The complete code should be obtained and read in conjunction with this book.

The aim of structural design is to produce a safe and economical structure that fulfils its required purpose. Theoretical knowledge of structural analysis must be combined with knowledge of design principles and theory and the constraints given in the standard to give a safe design. A thorough knowledge of properties of materials, methods of fabrication and erection is essential for the experienced designer. The learner must start with the basics and gradually build up experience through doing coursework exercises in conjunction with a study of design principles and theory.

British Standards are drawn up by panels of experts from the professional institutions, and include engineers from educational and research institutions, consulting engineers, government authorities and the fabrication and construction industries. The standards give the design methods, factors of safety, design loads, design strengths, deflection limits and safe construction practices.

As well as the main design standard for steelwork in buildings, BS 5950-1: 2000: Part 1, reference must be made to other relevant standards, including:

(1) BS EN 10020: 2000. This gives definition and classification of grades of steel.
(2) BS EN 10029: 1991 (plates); BS EN 10025: 1993 (sections); BS EN 10210-1: 1994 (hot finished hollow sections); BS EN 10219-1: 1997 (cold formed hollow sections). This gives the mechanical properties for the various types of steel sections.
(3) BS 6399-1: 1996 Part 1, Code of Practice for Dead and Imposed Loads.
(4) BS 6399-2: 1997 Part 2, Code of Practice for Wind Loads.
(5) BS 6399-3: 1998 Part 1, Code of Practice for Imposed Roof Loads.

Representative loading may be taken for element design. Wind loading depends on the complete building and must be estimated using the wind code.

1.5 Design calculations and computing

Calculations are needed in the design process to determine the loading on the structure, carry out the analysis and design the elements and joints, and must be set out clearly in a standard form. Design sketches to illustrate and amplify the calculations are an integral part of the procedure and are used to produce the detail drawings.

Computing now forms an increasingly larger part of design work, and all routine calculations can be readily carried out on a PC. The use of the computer speeds up calculation and enables alternative sections to be checked, giving the designer a wider choice than would be possible with manual working. However, it is most important that students understand the design principles involved before using computer programs.

It is through doing exercises that the student consolidates the design theory given in lectures. Problems are given at the end of most chapters.

1.6 Detailing

Chapter 12 deals with the detailing of structural steelwork. In the earlier chapters, sketches are made in design problems to show building arrangements, loading on frames, trusses, members, connections and other features pertinent to the design. It is often necessary to make a sketch showing the arrangement of a joint before the design can be carried out. At the end of the problem, sketches are made to show basic design information such as section size, span, plate sizes, drilling, welding, etc. These sketches are used to produce the working drawings.

The general arrangement drawing and marking plans give the information for erection. The detailed drawings show all the particulars for fabrication of the elements. The designer must know the conventions for making steelwork drawings, such as the scales to be used, the methods for specifying members, plates, bolts, welding, etc. He/she must be able to draw standard joint details and must also have a knowledge of methods of fabrication and erection. AutoCAD is becoming generally available and the student should be given an appreciation of their use.

2

Limit state design

2.1 Limit state design principles

The central concepts of limit state design are as follows:

(1) All the separate conditions that make the structure unfit for use are taken into account. These are the separate limit states.
(2) The design is based on the actual behaviour of materials and performance of structures and members in service.
(3) Ideally, design should be based on statistical methods with a small probability of the structure reaching a limit state.

The three concepts are examined in more detail below.

Requirement (1) means that the structure should not overturn under applied loads and its members and joints should be strong enough to carry the forces to which they are subjected. In addition, other conditions such as excessive deflection of beams or unacceptable vibration, though not in fact causing collapse, should not make the structure unfit for use.

In concept (2), the strengths are calculated using plastic theory and postbuckling behaviour is taken into account. The effect of imperfections on design strength is also included. It is recognized that calculations cannot be made in all cases to ensure that limit states are not reached. In cases such as brittle fracture, good practice must be followed to ensure that damage or failure does not occur.

Concept (3) implies recognition of the fact that loads and material strengths vary, approximations are used in design and imperfections in fabrication and erection affect the strength in service. All these factors can only be realistically assessed in statistical terms. However, it is not yet possible to adopt a complete probability basis for design, and the method adopted is to ensure safety by using suitable factors. Partial factors of safety are introduced to take account of all the uncertainties in loads, materials strengths, etc. mentioned above. These are discussed more fully below.

2.2 Limit states for steel design

The limit states for which steelwork is to be designed are set out in Section 2 of BS 5950-1: 2000. These are as follows.

8

2.2.1 Ultimate limit states

The ultimate limit states include the following:

(1) strength (including general yielding, rupture, buckling and transformation into a mechanism);
(2) stability against overturning and sway;
(3) fracture due to fatigue;
(4) brittle fracture.

When the ultimate limit states are exceeded, the whole structure or part of it collapses.

2.2.2 Serviceability limit states

The serviceability limit states consist of the following:

(5) deflection;
(6) vibration (for example, wind-induced oscillation);
(7) repairable damage due to fatigue;
(8) corrosion and durability.

The serviceability limit states, when exceeded, make the structure or part of it unfit for normal use but do not indicate that collapse has occurred.

All relevant limit states should be considered, but usually it will be appropriate to design on the basis of strength and stability at ultimate loading and then check that deflection is not excessive under serviceability loading. Some recommendations regarding the other limit states will be noted when appropriate, but detailed treatment of these topics is outside the scope of this book.

2.3 Working and factored loads

2.3.1 Working loads

The working loads (also known as the specified, characteristic or nominal loads) are the actual loads the structure is designed to carry. These are normally thought of as the maximum loads which will not be exceeded during the life of the structure. In statistical terms, characteristic loads have a 95 per cent probability of not being exceeded. The main loads on buildings may be classified as:

(1) *Dead loads*: These are due to the weights of floor slabs, roofs, walls, ceilings, partitions, finishes, services and self-weight of steel. When sizes are known, dead loads can be calculated from weights of materials or from the manufacturer's literature. However, at the start of a design, sizes are not known accurately and dead loads must often be estimated from experience. The values used should be checked when the final design is complete. For examples on element design, representative loading has been chosen, but for the building design examples actual loads from BS 6399: Part 1 are used.
(2) *Imposed loads*: These take account of the loads caused by people, furniture, equipment, stock, etc. on the floors of buildings and snow on roofs. The values of the floor loads used depend on the use of the building. Imposed loads are given in BS 6399: Part 1 and snow load is given in BS 6399: Part 3.

(3) *Wind loads*: These loads depend on the location and building size. Wind loads are given in BS 6399: Part 2.[*] Calculation of wind loads is given in the examples on building design.

(4) *Dynamic loads*: These are caused mainly by cranes. An allowance is made for impact by increasing the static vertical loads and the inertia effects are taken into account by applying a proportion of the vertical loads as horizontal loads. Dynamic loads from cranes are given in BS 6399: Part 1. Design examples show how these loads are calculated and applied to crane girders and columns.

Other loads on the structures are caused by waves, ice, seismic effects, etc. and these are outside the scope of this book.

2.3.2 Factored loads for the ultimate limit states

In accordance with Section 2.4.1 of BS 5950-1: 2000, factored loads are used in design calculations for strength and stability.

$$\text{Factored load} = \text{working or nominal load} \times \text{relevant partial load factor, } \gamma_f$$

The partial load factor takes account of:

(1) the unfavourable deviation of loads from their nominal values; and
(2) the reduced probability that various loads will all be at their nominal value simultaneously.

It also allows for the uncertainties in the behaviour of materials and of the structure as opposed to those assumed in design.

The partial load factors, γ_f are given in Table 2 of BS 5950-1: 2000 and some of the factors are given in Table 2.1.

Clause 2.4.1.1 of BS 5950-1: 2000 states that the factored loads should be applied in the most unfavourable manner and members and connections should

Table 2.1 Partial factors for load, γ_f

Loading	Factors γ_f
Dead load	1.4
Dead load restraining uplift or overturning	1.0
Dead load, wind load and imposed load	1.2
Imposed load	1.6
Wind load	1.4
Crane loads	
Vertical load	1.6
Vertical and horizontal load	1.4
Horizontal load	1.6
Crane loads and wind load	1.2

[*]*Note*: In countries other than United Kingdom, loads can be determined in accordance with this clause, or in accordance with local or national provisions as appropriate.

not fail under these load conditions. Brief comments are given on some of the load combinations:

(1) The main load for design of most members and structures is dead plus imposed load.
(2) In light roof structures uplift and load reversal occurs and tall structures must be checked for overturning. The load combination of dead plus wind load is used in these cases with a load factor of 1.0 for dead and 1.4 for wind load.
(3) It is improbable that wind and imposed loads will simultaneously reach their maximum values and load factors are reduced accordingly.
(4) It is also unlikely that the impact and surge load from cranes will reach maximum values together and so the load factors are reduced. Again, when wind is considered with crane loads the factors are further reduced.

2.4 Stability limit states

To ensure stability, Clause 2.4.2 of BS 5950 states that structures must be checked using factored loads for the following two conditions:

(1) *Overturning*: The structure must not overturn or lift off its seat.
(2) *Sway*: To ensure adequate resistance, two design checks are required:
    (a) Design to resist the applied horizontal loads.
    (b) A separate design for notional horizontal loads. These are to be taken as 0.5 per cent of the factored dead plus imposed load, and are to be applied at the roof and each floor level. They are to act with 1.4 times the dead and 1.6 times the imposed load.

Sway resistance may be provided by bracing rigid-construction shear walls, stair wells or lift shafts. The designer should clearly indicate the system he is using. In examples in this book, stability against sway will be ensured by bracing and rigid portal action.

2.5 Structural integrity

The provisions of Section 2.4.5 of BS 5950 ensure that the structure complies with the Building Regulations and has the ability to resist progressive collapse following accidental damage. The main parts of the clause are summarized below:

(1) All structures must be effectively tied at all floors and roofs. Columns must be anchored in two directions approximately at right angles. The ties may be steel beams or reinforcement in slabs. End connections must be able to resist a factored tensile load of 75 kN for floors and for roofs except where the steelwork only supports cladding that weighs not more than $0.7 \, \text{kN/m}^2$ and that carries only imposed roof loads and wind loads.
(2) Additional requirements are set out for certain multi-storey buildings where the extent of accidental damage must be limited. In general, tied buildings will be satisfactory if the following five conditions are met:
    (a) sway resistance is distributed throughout the building;
    (b) extra tying is to be provided as specified;

(c) column splices are designed to resist a specified tensile force;
(d) any beam carrying a column is checked as set out in (3) below; and
(e) precast floor units are tied and anchored.
(3) Where required in (2) the above damage must be localized by checking to see if at any storey any single column or beam carrying a column may be removed without causing more than a limited amount of damage. If the removal of a member causes more than the permissible limit, it must be designed as a key element. These critical members are designed for accidental loads set out in the Building Regulations.

The complete section in the code and the Building Regulations should be consulted.

2.6 Serviceability limit state deflection

Deflection is the main serviceability limit state that must be considered in design. The limit state of vibration is outside the scope of this book and fatigue was briefly discussed in Section 2.2.1 and, again, is not covered in detail. The protection for steel to prevent the limit state of corrosion being reached was mentioned in Section 2.2.4.

BS 5950-1: 2000 states in Clause 2.5.1 that deflection under serviceability loads of a building or part should not impair the strength or efficiency of the structure or its components or cause damage to the finishings. The serviceability loads used are the unfactored imposed loads except in the following cases:

(1) Dead + imposed + wind. Apply 80 per cent of the imposed and wind load.
(2) Crane surge + wind. The greater effect of either only is considered.

The structure is considered to be elastic and the most adverse combination of loads is assumed. Deflection limitations are given in Table 8 of BS 5950-1: 2000. These are given here in Table 2.2. These limitations cover beams and structures other than pitched-roof portal frames.

It should be noted that calculated deflections are seldom realized in the finished structure. The deflection is based on the beam or frame steel section only and composite action with slabs or sheeting is ignored. Again, the full value of the imposed load used in the calculations is rarely achieved in practice.

Table 2.2 Deflection limits

Deflection of beams due to unfactored imposed loads	
Cantilevers	Length/180
Beams carrying plaster	Span/360
All other beams	Span/200
Horizontal deflection of columns due to unfactored imposed and wind loads	
Tops of columns in single-storey buildings	Height/300
In each storey of a building with more than one storey	Storey height/300
Crane gantry girders	
Vertical deflection due to static wheel loads	Span/600
Horizontal deflection (calculated on top flange properties alone) due to crane surge	Span/500

2.7 Design strength of materials

The design strengths for steel complying with BS 5950-2 are given in Section 3.1.1 of BS 5950-1: 2000. Note that the material strength factor γ_m, part of the overall safety factor in limit state design, is taken as 1.0 in the code. The design strength may be taken as

$$p_y = 1.0\,Y_s \text{ but not greater than } U_s/1.2$$

where Y_s is the minimum yield strength, R_{eH} and U_s the minimum ultimate tensile strength, R_m.

For the common types of steel values of p_y are given in Table 9 of the code and reproduced in Table 2.3.

Table 2.3 Design strengths p_y for steel

Steel grade	Thickness (mm) less than or equal to	Sections, plates and hollow sections, p_y (N/mm^2)
S275	16	275
	40	265
	63	255
	80	245
	100	235
	150	225
S355	16	355
	40	345
	63	335
	80	325
	100	315
	150	295
S460	16	460
	40	440
	63	430
	80	410
	100	400

The code states that the following values for the elastic properties are to be used:

Modulus of elasticity, $E = 205\,000\,\text{N/mm}^2$
Shear modulus, $G = E/[2(1 + v)]$
Poisson's ratio, $v = 0.30$
Coefficient of linear thermal expansion
(in the ambient temperature range), $\alpha = 12 \times 10^{-6}/°C$

2.8 Design methods for buildings

The design of buildings must be carried out in accordance with one of the methods given in Clause 2.1.2 of BS 5950-1: 2000. The design methods are as follows:

(1) *Simple design*: In this method, the connections between members are assumed not to develop moments adversely affecting either the members or structure as a whole. The structure is assumed to be pin jointed for analysis. Bracing or shear walls are necessary to provide resistance to horizontal loading.

(2) *Continuous design*: The connections are assumed to be capable of developing the strength and/or stiffness required by an analysis assuming full continuity. The analysis may be made using either elastic or plastic methods.

(3) *Semi-continuous design*: This method may be used where the joints have some degree of strength and stiffness, but insufficient to develop full continuity. Either elastic or plastic analysis may be used. The moment capacity, rotational stiffness and rotation capacity of the joints should be based on experimental evidence. This may permit some limited plasticity, provided that the capacity of the bolts or welds is not the failure criterion. On this basis, the design should satisfy the strength, stiffness and in-plane stability requirements of all parts of the structure when partial continuity at the joints is taken into account in determining the moments and forces in the members.

(4) *Experimental verification*: The code states that where the design of a structure or element by calculation in accordance with any of the above methods is not practicable, the strength and stiffness may be confirmed by loading tests. The test procedure is set out in Section 7 of the code.

In practice, structures are designed to either the simple or continuous methods of design. Semi-continuous design has never found general favour with designers. Examples in this book are generally of the simple method of design.

3
Materials

3.1 Structural steel properties

Structural steel products are manufactured to conform to new specifications given in BS 5950 Part 2: 2001. The previously used specification for weldable structural steels, BS 4360: 1990 has been replaced by a series of Euronorm specifications for technical delivery requirements, dimensions and tolerances such as BS EN10025, BS EN10029, BS EN10051, BS EN10113, BS EN10137, BS EN10155, BS EN10163, BS EN10210, BS EN10219 and others.

Steel is composed of about 98 per cent of iron with the main alloying elements carbon, silicon and manganese. Copper and chromium are added to produce the weather-resistant steels that do not require corrosion protection. Structural steel is basically produced in three strength grades S275, S355 and S460. The important design properties are strength, ductility, impact resistance and weldability.

The stress–strain curves for the three grades of steel are shown in Figure 3.1(a) and these are the basis for the design methods used for steel. Elastic design is kept within the elastic region and because steel is almost perfectly elastic, design based on elastic theory is a very good method to use.

The stress–strain curves show a small plateau beyond the elastic limit and then an increase in strength due to strain hardening. Plastic design is based on the horizontal part of the stress–strain shown in Figure 3.1(b).

The mechanical properties for steels are set out in the respective specifications mentioned earlier. The yield strengths for the various grades vary with the thickness and other important design properties are given in Section 2.7 of this book.

3.2 Design considerations

Special problems occur with steelwork and good practice must be followed to ensure satisfactory performance in service. These factors are discussed briefly below in order to bring them to the attention of students and designers, although they are not generally of great importance in the design problems covered in this book. However, it is worth noting that the material safety factor γ_m is set to unity in BS 5950 which implies a certain level of quality and testing in steel usage. Weld procedures are qualified by maximum carbon equivalent values.

(a) Stress–strain diagrams for structural steels

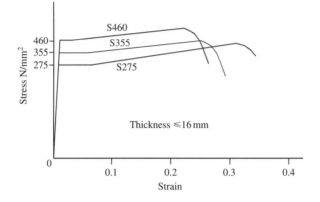

(b) Stress–strain diagram for plastic design

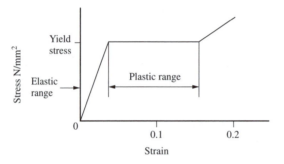

Figure 3.1 Stress–strain diagrams for structural steels

Attention to weldability should be given when dealing with special, thick and higher grade steel to avoid hydrogen induced cracking. Reader can refer to BS EN10229: 1998 for more information if necessary.

3.2.1 Fatigue

Fatigue failure can occur in members or structures subjected to fluctuating loads such as crane girders, bridges and offshore structures. Failure occurs through initiation and propagation of a crack that starts at a fault or structural discontinuity and the failure load may be well below its static value.

Welded connections have the greatest effect on the fatigue strength of steel structures. Tests show that butt welds give the best performance in service while continuous fillet welds are much superior to intermittent fillet welds. Bolted connections do not reduce the strength under fatigue loading. To help avoid fatigue failure, detail should be such that stress concentrations and abrupt changes of section are avoided in regions of tensile stress. Cases where fatigue could occur are noted in this book, and for further information the reader should consult reference (1).

3.2.2 Brittle fracture

Structural steel is ductile at temperatures above 10°C but it becomes more brittle as the temperature falls, and fracture can occur at low stresses below 0°C. The Charpy impact test is used to determine the resistance of steel to brittle fracture. In this test, a small specimen is broken by a hammer and the energy or toughness to cause failure at a given test temperature is measured.

In design, brittle fracture should be avoided by using steel quality grade with adequate impact toughness. Quality steels are designated JR, J0, J2, K2 and so forth in order of increasing resistance to brittle fracture. The Charpy impact fracture toughness is specified for the various steel quality grades: for example, Grade S275 J0 steel is to have a minimum fracture toughness of 27 J at a test temperature of 0°C.

In addition to taking care in the selection of steel grade to be used, it is also necessary to pay special attention to the design details to reduce the likelihood of brittle fracture. Thin plates are more resistant than thick ones. Abrupt changes of section and stress concentration should be avoided. Fillets welds should not be laid down across tension flanges and intermittent welding should not be used.

Cases where brittle fracture may occur in design of structural elements are noted in this book. For further information, the reader should consult reference (2).

3.2.3 Fire protection

Structural steelwork performs badly in fires, with the strength decreasing with increase in temperature. At 550°C, the yield stress has fallen to approximately 0.7 of its value at normal temperatures; that is, it has reached its working stress and failure occurs under working loads.

The statutory requirements for fire protection are usually set out clearly in the approved documents from the local Building Regulations (3) or Fire Safety Authority. These lay down the fire-resistance period that any load-bearing element in a given building must have, and also give the fire-resistance periods for different types of fire protection. Fire protection can be provided by encasing the member in concrete, fire board or cementitious fibre materials. The main types of fire protection for columns and beams are shown in Figure 3.2. More recently, intumescent paint is being used especially for exposed steelwork.

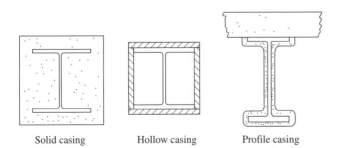

Solid casing   Hollow casing   Profile casing

Figure 3.2 Fire protections for columns and beams

All multi-storey steel buildings require fire protection. Single-storey factory buildings normally do not require fire protection for the steel frame. Further information is given in reference (4).

3.2.4 Corrosion protection

Exposed steelwork can be severely affected by corrosion in the atmosphere, particularly if pollutants are present, and it is necessary to provide surface protection in all cases. The type of protection depends on the surface conditions and length of life required.

The main types of protective coatings are:

(1) *Metallic coatings*: Either a sprayed-on in line coating of aluminium or zinc is used or the member is coated by hot-dipping it in a bath of molten zinc in the galvanizing process.
(2) *Painting*: where various systems are used. One common system consists of using a primer of zinc chromate followed by finishing coats of micaceous iron oxide. Plastic and bituminous paints are used in special cases.

The single most important factor in achieving a sound corrosion-protection coating is surface preparation. Steel is covered with mill scale when it cools after rolling, and this must be removed before the protection is applied, otherwise the scale can subsequently loosen and break the film. Blast cleaning makes the best preparation prior to painting. Acid pickling is used in the galvanizing process. Other methods of corrosion protection which can also be considered are sacrificial allowance, sherardizing, concrete encasement and cathodic protection.

Careful attention to design detail is also required (for example, upturned channels that form a cavity where water can collect should be avoided) and access for future maintenance should also be provided. For further information the reader should consult BS EN ISO12944 Parts 1-8: 1998—Corrosion Protection of Steel Structures by Protective Paint Systems and BS EN ISO14713: 1999—Protection against Corrosion of Iron and Steel in Structures, Zinc and Aluminium Coatings.

3.3 Steel sections

3.3.1 Rolled and formed sections

Rolled and formed sections are produced in steel mills from steel blooms, beam blanks or coils by passing them through a series of rollers. The main sections are shown on Figure 3.3 and their principal properties and uses are discussed briefly below:

(1) *Universal beams*: These are very efficient sections for resisting bending moment about the major axis.
(2) *Universal columns*: These are sections produced primarily to resist axial load with a high radius of gyration about the minor axis to prevent buckling in that plane.
(3) *Channels*: These are used for beams, bracing members, truss members and in compound members.

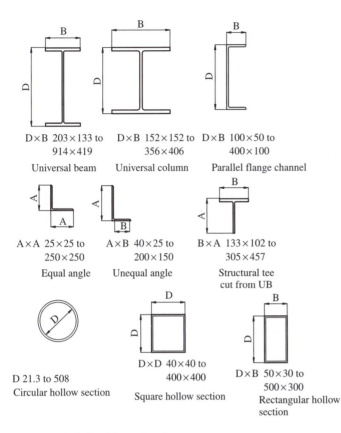

Figure 3.3 Rolled and formed sections

(4) *Equal and unequal angles*: These are used for bracing members, truss members and for purlins, side and sheeting rails.

(5) *Structural tees*: The sections shown are produced by cutting a universal beam or column into two parts. Tees are used for truss members, ties and light beams.

(6) *Circular, square and rectangular hollow sections*: These are mostly produced from hot-rolled coils, and may be hot-finished or cold-formed. A welded mother tube is first formed and then it is rolled to its final square or rectangular shape. In the hot process, the final shaping is done at the steel normalising temperature whereas in the cold process, it is done at ambient room temperature. Both types of hollow sections are now permitted in BS 5950. These sections make very efficient compression members, and are used in a wide range of applications as members in roof trusses, lattice girders, in building frames, for purlins, sheeting rails, etc.

Note that the range in serial sizes is given for the members shown in Figure 3.3. A number of different members are produced in each serial size by varying the flange, web, leg or wall thicknesses. The material properties, tolerances and dimensions of the various sections can be found in the following standards as shown in Table 3.1.

Table 3.1 Material properties, dimensions and tolerances of various sections

Sections	Materials	Dimensions and tolerances
Universal beams, columns, tees, bearing piles	EN 10025 EN 10113	EN 10034 BS 4-1
Channels (hot-finished)	EN 10025	BS 4-1
Purlins (cold-formed)	EN 10149	BS 5950-7
Angles	EN 10025	EN 10056
Flats (strips)	EN 10025	EN 10048
Plates	EN 10025	EN 10029
Hot-finished hollows	EN 10210-1	EN 10210-2
Cold-formed hollows	EN 10219-1	EN 10219-2

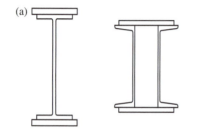

(a)

Compound beam

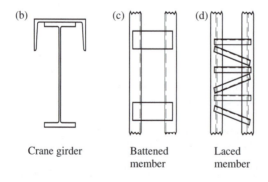

(b)   (c)   (d)

Crane girder   Battened   Laced
                    member   member

Figure 3.4 Compound sections

3.3.2 Compound sections

Compound sections are formed by the following means (Figure 3.4):

(1) strengthening a rolled section such as a universal beam by welding on cover plates, as shown in Figure 3.4(a);
(2) combining two separate rolled sections, as in the case of the crane girder in Figure 3.4(b). The two members carry loads from separate directions.
(3) connecting two members together to form a strong combined member. Examples are the laced and battened members shown in Figures 3.4(c) and (d).

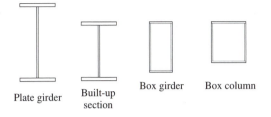

Plate girder   Built-up section   Box girder   Box column

Figure 3.5 Built-up sections

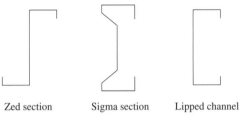

Zed section   Sigma section   Lipped channel

Figure 3.6 Cold-rolled sections

3.3.3 Built-up sections

Built-up sections are made by welding plates together to form I, H or box members which are termed plate girders, built-up columns, box girders or columns, respectively. These members are used where heavy loads have to be carried and in the case of plate and box girders where long spans may be required. Examples of built-up sections are shown in Figure 3.5.

3.3.4 Cold-rolled open sections

Thin steel plates can be formed into a wide range of sections by cold rolling. The most important uses for cold-rolled open sections in steel structures are for purlins, side and sheeting rails. Three common sections-the zed, sigma and lipped channel-are shown in Figure 3.6. Reference should be made to manufacturer's specialised literature for the full range of sizes available and the section properties. Some members and their properties are given in Sections 4.12.6 and 4.13.5 in design of purlins and sheeting rails.

3.4 Section properties

For a given member serial size, the section properties are:

(1) the exact section dimensions;
(2) the location of the centroid if the section is asymmetrical about one or both axes;
(3) area of cross-section;
(4) moments of inertia about various axes;

(5) radii of gyration about various axes;
(6) moduli of section for various axes, both elastic and plastic.

The section properties for hot rolled and formed sections are also listed in SCI Publication 202: *Steelwork Design Guide to BS 5950: Part 1: 2000, Volume 1 Section Properties and Member Capacities*, 6th edition with amendments, The Steel Construction Institute, United Kingdom.

For compound and built-up sections, the properties must be calculated from first principles. The section properties for the symmetrical I-section with dimensions as shown in Figure 3.7(a) are as follows:

(1) *Elastic properties*:

Area
$$A = 2BT + dt$$
Moment of inertia x–x axis
$$I_x = BD^3/12 - (B - t)d^3/12$$
Moment of inertia y–y axis
$$I_y = 2TB^3/12 + dt^3/12$$
Radius of gyration x–x axis
$$r_x = (I_x/A)^{0.5}$$
Radius of gyration y–y axis
$$r_y = (I_y/A)^{0.5}$$
Modulus of section x–x axis
$$Z_x = 2I_x/D$$
Modulus of section y–y axis
$$Z_y = 2I_y/B$$

(a)

Symmetrical I-section

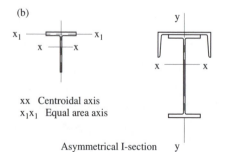

(b)

xx   Centroidal axis
x_1x_1 Equal area axis

Asymmetrical I-section

Figure 3.7 Beam section

(2) *Plastic moduli of section*: The plastic modulus of section is equal to the algebraic sum of the first moments of area about the equal area axis. For the I-section shown:

$$S_x = 2BT(D - T)/2 + td^2/4$$
$$S_y = 2TB^2/4 + dt^2/4$$

For asymmetrical sections such as those shown in Figure 3.7(b), the neutral axis must be located first. In elastic analysis, the neutral axis is the centroidal axis while in plastic analysis it is the equal area axis. The other properties may then be calculated using procedures from strength of materials (5). Calculations of properties for unsymmetrical sections are given in various parts of this book.

Other properties of universal beams, columns, joists and channels, used for determining the buckling resistance moment are:

buckling parameter, u;
torsional index, x;
warping constant, H;
torsional constant, J.

These properties may be calculated from formulae given in Appendix B of BS 5950: Part I or from Section A of the SCI Publication 202: *Steelwork Design Guide to BS 5950: Part 1: 2000, Volume 1 Section Properties and Member Capacities*, 6th edition with amendments, The Steel Construction Institute, United Kingdom.

4

Beams

4.1 Types and uses

Beams span between supports to carry lateral loads which are resisted by
bending and shear. However, deflections and local stresses are also important.
Beams may be cantilevered, simply supported, fixed ended or continuous, as
shown in Figure 4.1(a). The main uses of beams are to support floors and
columns, carry roof sheeting as purlins and side cladding as sheeting rails.

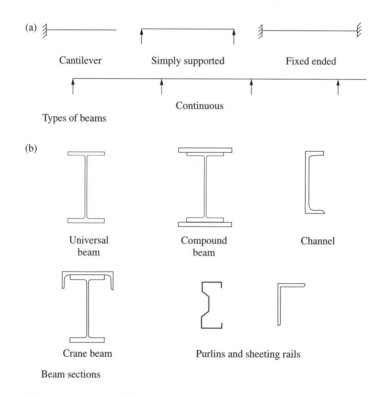

Figure 4.1 Types of beams and beam sections

24

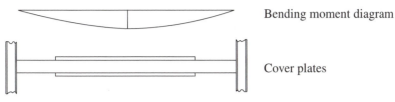

Bending moment diagram

Cover plates

Simply supported beam

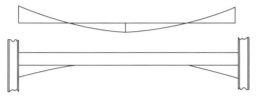

Bending moment diagram

Haunched ends

Fixed ended beam

Figure 4.2 Non-uniform beam

Any section may serve as a beam, and common beam sections are shown in Figure 4.1(b). Some comments on the different sections are given:

(1) The universal beam where the material is concentrated in the flanges is the most efficient section to resist uniaxial bending.
(2) The universal column may be used where the depth is limited, but it is less efficient.
(3) The compound beam consisting of a universal beam and flange plates is used where the depth is limited and the universal beam itself is not strong enough to carry the load.
(4) The crane beam consists of a universal beam and channel. It is because the beam needs to resist bending in both horizontal and vertical directions.

Beams may be of uniform or non-uniform section. Sections may be strengthened in regions of maximum moment by adding cover plates or haunches. Some examples are shown in Figure 4.2.

4.2 Beam loads

Types of beam loads are:

(1) concentrated loads from secondary beams and columns;
(2) distributed loads from self-weight and floor slabs.

The loads are further classified into:

(1) dead loads from self weight, slabs, finishes, etc.
(2) imposed loads from people, fittings, snow on roofs, etc.
(3) wind loads, mainly on purlins and sheeting rails.

Loads on floor beams in a steel frame building are shown in Figure 4.3(a). The figure shows loads from a two-way spanning slab which gives trapezoidal

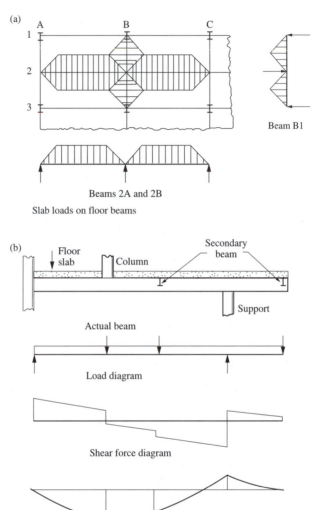

Beam B1

Beams 2A and 2B

Slab loads on floor beams

Actual beam

Load diagram

Shear force diagram

Bending moment diagram

Actual loads on a beam

Figure 4.3 Beam loads

and triangular loads on the beams. One-way spanning floor slabs give uniform loads. An actual beam with the floor slab and members it supports is shown in Figure 4.3(b). The load diagram and shear force and bending moment diagrams constructed from it are also shown.

4.3 Classification of beam cross-sections

The projecting flange of an I-beam will buckle prematurely if it is too thin. Webs will also buckle under compressive stress from bending and from shear.

This problem is discussed in more detail in Section 5.2 in Chapter 5 (see also reference (6)).

To prevent local buckling from occurring, limiting outstand/thickness ratios for flanges and depth/thickness ratios for webs are given in BS 5950-1: 2000 in Section 3.5. Beam cross-sections are classified as follows in accordance with their behaviour in bending:

> *Class 1 Plastic cross-section*: This can develop a plastic hinge with sufficient rotation capacity to permit redistribution of moments in the structure. Only class I sections can be used for plastic design.
> *Class 2 Compact cross-section*: This can develop the plastic moment capacity, but local buckling prevents rotation at constant moment.
> *Class 3 Semi-compact cross-section*: The stress in the extreme fibres should be limited to the yield stress because local buckling prevents development of the plastic moment capacity.
> *Class 4 Slender cross-section*: Premature buckling occurs before yield is reached.

Flat elements in a cross section are classified as:

(1) Internal elements supported on both longitudinal edges.
(2) Outstand elements attached on one edge with the other free.

Elements are generally of uniform thickness, but, if tapered, the average thickness is used. Elements are classified as plastic, compact or semi-compact if they meet limits given in Tables 11 and 12 in association with Figures 5 and 6 of the code. An example for the limiting proportions for elements of universal beams and channels are shown in Figure 4.4.

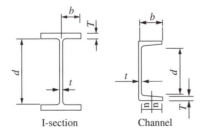

I-section   Channel

Compression element		Ratio	Limiting value		
			Class 1 plastic	Class 2 compact	Class 3 semi-compact
Outstand element of compression flange	Rolled section	b/T	9ε	10ε	15ε
Web with neutral axis at mid-depth		d/t	80ε	100ε	120ε

The parameter, $\varepsilon = (275/p_y)^{0.5}$

Figure 4.4 Limiting proportions for rolled sections

4.4 Bending stresses and moment capacity

Both elastic and plastic theories are discussed here. Short or restrained beams are considered in this section. Plastic properties are used for plastic and compact sections and elastic properties for semi-compact sections to determine moment capacities. For slender sections, only effective elastic properties are used.

4.4.1 Elastic theory

(1) Uniaxial bending

The bending stress distributions for an I-section beam subjected to uniaxial moment are shown in Figure 4.5(a). We define following terms for the I-section:

$M =$ applied bending moment;
$I_x =$ moment of inertia about $x–x$ axis;
$Z_x = 2I_x/D =$ modulus of section for $x–x$ axis; and
$D =$ overall depth of beam.

The maximum stress in the extreme fibres top and bottom is:

$$f_{bc} = f_{bt} = M_x/Z_x$$

The moment capacity, $M_c = \sigma_b Z_x$ where σ_b is the allowable stress.

The moment capacity for a semi-compact section subjected to a moment due to factored loads is given in Clause 4.2.5.2 of BS 5950-1: 2000 as

$$M_c = p_y Z$$

where p_y is the design strength.

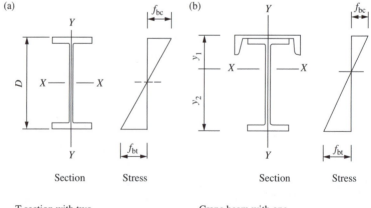

Section   Stress   Section   Stress

T-section with two   Crane beam with one
axes of symmetry   axis of symmetry

Figure 4.5 Beams in uniaxial bending

For the asymmetrical crane beam section shown in Figure 4.5(b), the additional terms require definition:

$Z_1 = I_x/y_1 =$ modulus of section for top flange,
$Z_2 = I_x/y_2 =$ modulus of section for bottom flange,
$y_1, y_2 =$ distance from centroid to top and bottom fibres.

The bending stresses are:

Top fibre in compression $f_{bc} = M_x/Z_1$
Bottom fibre in tension $f_{bt} = M_x/Z_2$

The moment capacity controlled by the stress in the bottom flange is

$$M_c = p_y Z_2$$

(2) Biaxial bending

Consider that I-section in Figure 4.6(a) which is subject to bending about both axes. We define the following terms:

$M_x =$ moment about the x–x axis,
$M_y =$ moment about the y–y axis,
$Z_x =$ modulus of section for the x–x axis,
$Z_y =$ modulus of section for the y–y axis.

The maximum stress at A or B is:

$$f_A = f_B = M_x/Z_x + M_y/Z_y$$

If the allowable stress is σ_b, the moment capacities with respect to x–x and y–y axes are:

$$M_{cx} = \sigma_b Z_x$$
$$M_{cy} = \sigma_b Z_y$$

Taking the maximum stress as σ_b and substituting for Z_x and Z_y in the expression above gives the interaction relationship

$$\frac{M_x}{M_{cx}} + \frac{M_y}{M_{cy}} = 1$$

This is shown graphically in Figure 4.6(b).

(3) Asymmetrical sections

Note that with the channel section shown in Figure 4.7(a), the vertical load must be applied through the shear centre for bending in the free member to

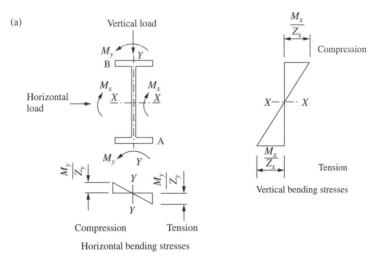

(a)

Bending stresses

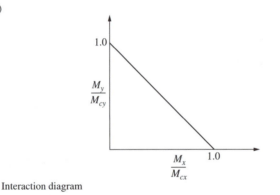

(b)

Interaction diagram

Figure 4.6 Biaxial bending

take place about the x–x axis, otherwise twisting and biaxial bending occurs. However, a horizontal load applied through the centroid causes bending about the y–y axis only.

For an asymmetrical section such as the unequal angle shown in Figure 4.7(b), bending takes place about the principal axes u–u and v–v in the free member when the load is applied through the shear centre. When the angle is used as a purlin, the cladding restrains the member so that it bends about the x–x axis.

4.4.2 Plastic theory

(1) Uniaxial bending

The stress–strain curve for steel on which plastic theory is based is shown in Figure 4.8(a). In the plastic region after yield, the strain increases without

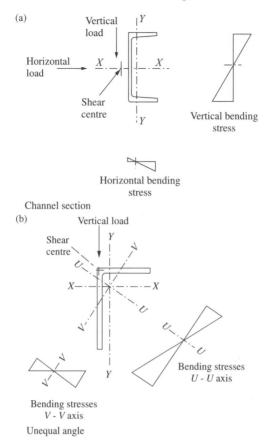

Figure 4.7 **Bending of asymmetrical sections**

increase in stress. Consider the I-section shown in Figure 4.8(b). Under moment, the stress first follows an elastic distribution. As the moment increases, the stress at the extreme fibre reaches the yield stress and the plastic region proceeds inwards as shown, until the full plastic moment is reached and a plastic hinge is formed.

For single axis bending, the following terms are defined:

M_c = plastic moment capacity,
S = plastic modulus of section,
Z = elastic modulus of section,
p_y = design strength.

The moment capacity given in Clause 4.2.5.2 of BS 5950-1: 2000 for class 1 and 2 sections with low shear load is:

$$M_c = p_y S,$$
$$\leq 1.2\, p_y Z.$$

(a)

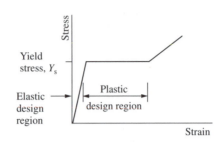

Simplified stress-strain curve

(b)

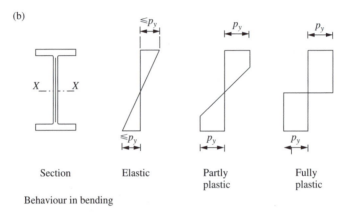

Figure 4.8 Behaviour in bending

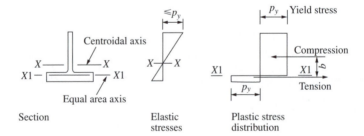

Figure 4.9 Section with one axis of symmetry

The first expression is the plastic moment capacity, the second ensures that yield does not occur at working loads in I-sections bent about the y–y axis.

For single-axis bending for a section with one axis of symmetry, consider the T-section shown in Figure 4.9. In the elastic range, bending takes place about the centroidal axis and there are two values for the elastic modulus of section.

In the plastic range, bending takes place about the equal area axis and there is one value for the plastic modulus of section:

$$S = M_c/p_y$$
$$= Ab/2$$

where A is the area of cross section and b is the lever arm between the tension and compression forces.

(2) Biaxial bending

When a beam section is bent about both axes, the neutral axis will lie at an angle to the rectangular axes which depends on the section properties and values of the moments. Solutions have been obtained for various cases and a relationship established between the ratios of the applied moments and the moment capacities about each axis. The relationship expressed in Sections 4.9 of BS 5950-1: 2000 for plastic or compact cross sections is given in the following form:

$$\left(\frac{M_x}{M_{cx}}\right)^{Z_1} + \left(\frac{M_y}{M_{cy}}\right)^{Z_2} \leq 1$$

where M_x = factored moment about the x–x axis,
M_y = factored moment about the y–y axis,
M_{cx} = moment capacity about the x–x axis,
M_{cy} = moment capacity about the y–y axis,
Z_1 = 2 for I- and H-sections and 1 for other open sections
Z_2 = 1 for all open sections.

A conservative result is given if $Z_1 = Z_2 = 1$. The interaction diagram is shown in Figure 4.10.

(3) Unsymmetrical sections

For sections with no axis of symmetry, plastic analysis for bending is complicated, but solutions have been obtained. In many cases where such sections are used, the member is constrained to bend about the rectangular axis (see Section 4.4.1(3)). Such cases can also be treated by elastic theory using factored loads with the maximum stress limited to the design strength.

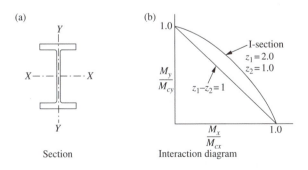

(a) Section

(b) Interaction diagram

Figure 4.10 Biaxial bending

4.5 Lateral torsional buckling

4.5.1 General considerations

The compression flange of an I-beam acts like a column, and will buckle sideways if the beam is not sufficiently stiff or the flange is not restrained laterally. The load at which the beam buckles can be much less than that causing the full moment capacity to develop. Only a general description of the phenomenon and factors affecting it are set out here. The reader should consult references (6) and (7) for further information.

Consider the simply supported beam with ends free to rotate in plan but restrained against torsion and subjected to end moments, as shown in Figure 4.11. Initially, the beam deflects in the vertical plane due to bending, but as the moment increases, it reaches a critical value M_E less than the moment capacity, where it buckles sideways, twists and collapses.

Elastic theory is used to set up equilibrium equations to equate the disturbing effect to the lateral bending and torsional resistances of the beam. The solution of this equation gives the elastic critical moment:

$$M_E = \frac{\pi}{L}\sqrt{EI_yGJ}\sqrt{1 + \frac{\pi^2EH}{L^2GJ}}$$

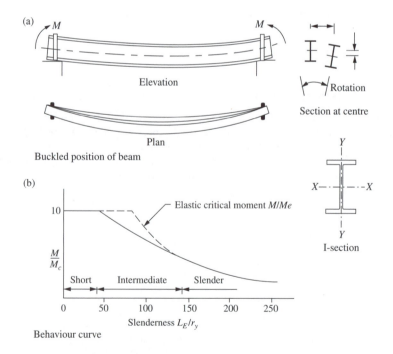

Figure 4.11 Lateral torsional buckling

where E = Young's modulus,
$\quad\quad G$ = shear modulus,
$\quad\quad J$ = torsion constant for the section,
$\quad\quad H$ = warping constant for the section,
$\quad\quad L$ = span
$\quad\quad I_y$ = moment of inertia about the y–y axis.

The theoretical solution applies to a beam subjected to a uniform moment. In other cases where the moment varies, the tendency to buckling is reduced. If the load is applied to the top flange and can move sideways, it is destabilizing, and buckling occurs at lower loads than if the load were applied at the centroid, or to the bottom flange.

In the theoretical analysis, the beam was assumed to be straight. Practical beams have initial curvature and twisting, residual stresses, and the loads are applied eccentrically. The theory set out above requires modification to cover actual behaviour. Theoretical studies and tests show that slender beams fail at the elastic critical moment, M_E and short or restrained beams fail at the plastic moment capacity M_c. A lower bound curve running between the two extremes can be drawn to contain the behaviour of intermediate beams. Beam behaviour as a function of slenderness is shown in Figure 4.11(b).

To summarize, factors influencing lateral torsional buckling are:

(1) The unrestrained length of compression flange: The longer this is, the weaker the beam. Lateral buckling is prevented by providing props at intermediate points.
(2) The end conditions: Rotational restraint in plan helps to prevent buckling.
(3) Section shape: Sections with greater lateral bending and torsional stiffness have greater resistance to buckling.
(4) Note that lateral restraint to the tension flange also helps to resist buckling (see Figure 4.11).
(5) The application of the loads and shape of the bending moment diagram between restraints.

A practical design procedure must take into account the effects noted above. Terms used in the curve are defined as follows:

$\quad M$ = moment causing failure,
$\quad M_c$ = moment capacity for a restrained beam,
$\quad M_E$ = elastic critical moment,
$\quad L_E/r_y$ = slenderness with respect to the y–y axis (see the next section).

4.5.2 Lateral restraints and effective length

The code states in Clause 4.2.2 that full lateral restraint is provided by a floor slab if the friction or shear connection is capable of resisting a lateral force of 2.5 per cent of the maximum factored force in the compression flange. Other suitable construction can also be used. Members not provided with full lateral restraint must be checked for buckling.

The following two types of restraints are defined in Sections 4.3.2 and 4.3.3 of the code:

(1) Intermediate lateral restraint, which prevents sideways movement of the compression flange; and

(a)

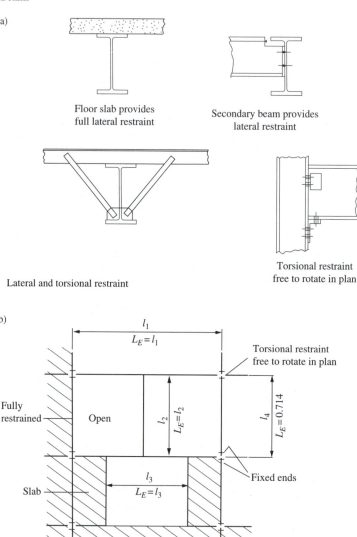

Figure 4.12 Restraints and effective lengths

(2) Torsional restraint, which prevents movement of one flange relative to the other.

Restraints are provided by floor slabs, end joints, secondary beams, stays, sheeting, etc., and some restraints are shown in Figure 4.12(a).

The effective length L_E for a beam is defined in Section 1 of the code as the length between points of effective restraints multiplied by a factor to take account of the end conditions and loading. Note that a destabilizing load (where the load is applied to the top flange and can move with it) is taken account of by increasing the effective length of member under consideration.

Table 4.1 Effective length L_E–Beams

Support conditions	Loading conditions	
	Normal	Destabilizing
Beam partial torsionally unrestrained Compression flange laterally unrestrained Both flanges free to rotate on plan	$1.2L_{LT} + 2D$	$1.4L_{LT} + 2D$
Beam torsionally restrained Compression flange laterally restrained Compression flange only free to rotate on plan	$1.0L_{LT}$	$1.2L_{LT}$
Beam torsionally restrained Both flanges NOT free to rotate on plan	$0.7L_{LT}$	$0.85L_{LT}$

L_{LT} = length of beam between restraints.
D = depth of beam.

The effective length for beams is discussed in Section 4.3.5 of BS 5950-1: 2000. When the beam is restrained at the ends only, that is, without intermediate restraint, the effective length should be obtained from Table 13 in the code. Some values from this table are given in Table 4.1.

Where the beam is restrained at intervals by other members the effective length L_E may be taken as L, the distance between restraints. Some effective lengths for floor beams are shown in Figure 4.12(b).

4.5.3 Code design procedure

(1) General procedure

The general procedure for checking the resistance to lateral torsional buckling is out lined in Section 4.3.6 of BS 5950-1: 2000:

(1) Resistance to lateral-torsional buckling need not be checked separately (and the buckling resistance moment M_b may be taken as equal to the relevant moment capacity M_c) in the following cases:

bending about the minor axis;
CHS, square RHS or circular or square solid bars;
RHS, unless L_E/r_y exceeds the limiting value given in Table 15 of BS 5950-1: 2000 for the relevant value of D/B;
I, H, channel or box sections, if λ_{LT} does not exceed λ_{L0},

Otherwise, for members subject to bending about their major axis, reference should be made as follows:

$$M_x \leq M_b/m_{LT} \quad \text{and} \quad M_x \leq M_{cx}$$

(2) Calculate the equivalent uniform moment factor m_{LT}:
The value of the equivalent uniform moment factor m_{LT} which depend on the ratio and direction of the major axis moment. For the normal loading condition, the equivalent uniform moment factor for lateral-torsional buckling should be obtained from Table 18 of BS 5950-1: 2000. For destabilizing loading condition, m_{LT} should be taken as 1.0. Values for some common load cases are shown in Figure 4.13.

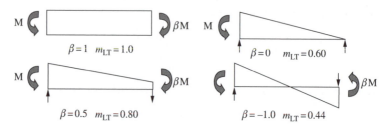

Figure 4.13 Equivalent uniform moment factor m_{LT} for lateral-torsional buckling

(3) Estimate the effective length L_E of the unrestrained compression flange using the rules from Section 4.5.2. Minor axis slenderness, $\lambda = L_E/r_y$, where r_y = radius of gyration for the y–y axis.

(4) Calculate the equivalent slenderness, λ_{LT}

$$\lambda_{LT} = uv\lambda\sqrt{\beta_w}$$

where

u = buckling parameter allowing for torsional resistance. This may be calculated from the formulae in Appendix B or taken from the published table in the *Guide to BS 5950: Part 1: 2000, Vol. 1, Section properties, Member Capacities, SCI*. It can also conservatively taken as 0.9 for an uniform rolled I-section,

v = slenderness factor which depends on values of η and λ/x.

where

$$\eta = \frac{I_{yc}}{I_{yc} + I_{yt}}$$

I_{yc} = second moment of area of the compression flange about the minor axis of the section;

I_{yt} = second moment of area of the tension flange about the minor axis of the section,

η = 0.5 for a symmetrical section,

x = torsional index. This can be calculated from the formula in Appendix B or obtained from the published table in the *Guide to BS 5950: Part 1: 2000*. The torsional index can be taken conservatively approximately equal to D/T where D is the overall depth of beam and T the thickness of the compression flange.

Values of v are given in Table 19 of the BS 5950-1: 2000. Alternatively, v can be determined by the formulae in Appendix B or Clause 4.3.6.7. in the code.

(5) Ratio β_w should be determined in accordance to Clause 4.3.6.9. The ratio is dependent on the classification of the sections. For class 1 or class 2 sections, β_w is taken as 1.0.

(6) Read the bending strength, p_b from Table 16 for rolled sections and Table 17 for welded sections in the BS 5950-1: 2000. Values of p_b depend on the equivalent slenderness λ_{LT} and design strength p_y.

(7) Calculate the buckling resistance moment.
for class 1 plastic or class 2 compact cross-sections:

$$M_b = p_b S_x.$$

for class 3 semi-compact cross-sections:

$$M_b = p_b Z_x; \quad \text{or alternatively,} \quad M_b = p_b S_{x,\text{eff}}$$

for class 4 slender cross-sections:

$$M_b = p_b Z_{x,\text{eff}}.$$

where

p_b is the bending strength;
S_x is the plastic modulus about the major axis;
$S_{x,\text{eff}}$ is the effective plastic modulus about the major axis;
Z_x is the section modulus about the major axis;
$Z_{x,\text{eff}}$ is the effective section modulus about the major axis.

(2) Conservative approach for equal flanged rolled sections

The code gives a conservative approach for equal flanged rolled sections in Section 4.3.7. The buckling resistance moment M_b of a plain rolled I, H or channel section with equal flanges may be determined using the bending strength, p_b obtained from Table 20 for the relevant values of $(\beta_w)^{0.5} L_E / r_y$ and D/T as follows:

for a class 1 plastic or class 2 compact cross-section:

$$M_b = p_b S_x$$

for a class 3 semi-compact cross-section:

$$M_b = p_b Z_x$$

4.5.4 Biaxial bending

Lateral torsional buckling affects the moment capacity with respect to the major axis only of I-section beams. When the section is bent about only the minor axis, it will reach the moment capacity given in Section 4.4.2(1).

Where biaxial bending occurs, BS 5950-1: 2000 specifies in Section 4.9 that the following simplified interaction expressions must be satisfied for plastic or compact sections:

(1) Cross-section capacity check at point of maximum combined moments:

$$\frac{M_x}{M_{cx}} + \frac{M_y}{M_{cy}} \leq 1$$

This design check was discussed in Section 4.4.2(2) above.

(2) Member buckling check at the centre of the beam:

$$\frac{m_x M_x}{p_y Z_x} + \frac{m_y M_y}{p_y Z_y} \le 1$$

$$\frac{m_{LT} M_{LT}}{M_b} + \frac{m_y M_y}{p_y Z_y} \le 1$$

where

M_b is the buckling resistance moment,
M_{LT} is the maximum major axis moment in the segment length L governing M_b;
M_x is the maximum major axis moment in the segment length L_x;
M_y is the maximum minor axis moment in the segment length L_y;
Z_x is the section modulus about the major axis;
Z_y is the section modulus about the minor axis.

The equivalent uniform moment factors m_{LT}, m_x and m_y should be obtained from Clause 4.8.3.3.4 of BS 5950-1: 2000.
   More exact expressions are given in the code. Biaxial bending is discussed more fully in Chapter 7 of this book.

4.6 Shear in beams

4.6.1 Elastic theory

The value of shear stress at any point in a beam section is given by the following expression (see Figure 4.14(a)):

$$f_s = \frac{V A y}{I_x t}$$

where V = shear force at the section
     A = area between the point where the shear stress is required and a free edge
     y = distance from the centroid of the area A to the centroid of the section
     I_x = second moment of area about the $x–x$ axis
     t = thickness of the section at the point where the shear stress is required.

   Using this formula, the shear stresses at various points in the beam section can be found. Thus, the maximum shear stress at the centroid in terms of the beam dimensions shown in the figure is:

$$f_{max} = \frac{V}{I_x t} \left(\frac{BT(d+T)}{2} + \frac{td^2}{8} \right)$$

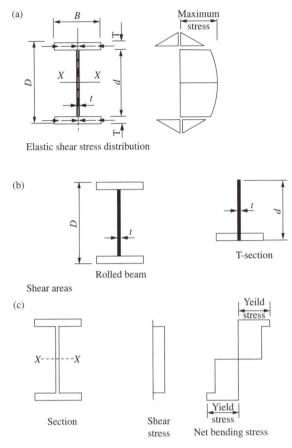

Figure 4.14 Shear in beams

Note that the distribution shows that the web carries the bulk of the shear. It has been customary in design to check the average shear stress in the web given by:

$$f_{av} = V/Dt$$

which should not exceed an allowable value.

4.6.2 Plastic theory

Shear is considered in BS 5950-1: 2000 in Section 4.2.3. For a rolled member subjected to shear only, the shear force is assumed to be resisted by the web area A_v shown in Figure 4.14(b), where:

$$A_v = \text{web thickness} \times \text{overall depth} = tD$$

For the T section shown in the figure:

$$A_v = 0.9A_o$$

where A_o is the area of the rectilinear element which has the largest dimension in the direction parallel to the shear force and equal to td.

The shear area may be stressed to the yield stress in shear, that is, to $1/\sqrt{3}$ of the yield stress in tension. The capacity is given in the code as:

$$P_v = 0.6p_y A_v$$

If the ratio d/t exceeds 70ε for a rolled section, or 62ε for a welded section, the web should be checked for shear buckling in accordance with Clause 4.4.5 in BS 5950-1: 2000.

If moment as well as shear occurs at the section, the web is assumed to resist all the shear while the flanges are stressed to yield by bending. The section analysis is based on the shear stress and bending stress distributions shown in Figure 4.14(c). The web is at yield under the combined bending and shear stresses and von Mises' criterion is adopted for failure in the web. The shear reduces the moment capacity, but the reduction is small for all but high values of shear force. The analysis for shear and bending is given in reference (8).

BS 5950-1: 2000 gives the following expression in Section 4.2.5.3 for the moment capacity for plastic or compact sections in the presence of high shear load.

When the average shear force F, is less than 0.6 of the shear capacity P_v, no reduction in moment capacity is required. When F_v is greater than 0.6 P_v, the reduced moment capacity for class 1 or class 2 cross-sections is given by:

$$M_c = p_y(S - \rho S_v)$$

where $S_v = tD^2/4$ for a rolled section with equal flanges
$$\rho = [2(F_v/P_v) - 1]^2.$$

4.7 Deflection of beams

The deflection limits for beams specified in Section 2.5.1 of BS 5950-1: 2000 were set in Section 2.6 of this book. The serviceability loads are the unfactored imposed loads.

Deflection formulae are given in design manuals. Deflections for some common load cases for simply supported beams together with the maximum moments are given in Figure 4.15. For general load cases deflections can be calculated by the moment area method. (see references (9) and (10).)

4.8 Beam connections

End connections to columns and other beams form an essential part of beam design. Checks for local failure are required at supports and points where concentrated loads are applied.

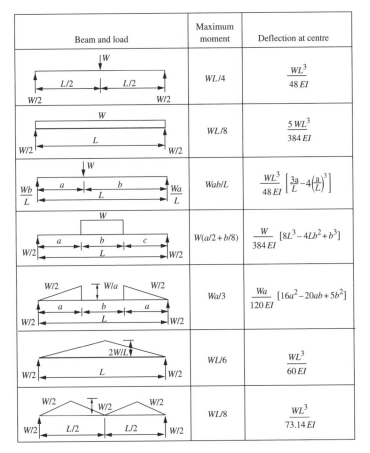

Beam and load	Maximum moment	Deflection at centre
	$WL/4$	$\dfrac{WL^3}{48\,EI}$
	$WL/8$	$\dfrac{5\,WL^3}{384\,EI}$
	Wab/L	$\dfrac{WL^3}{48\,EI}\left[\dfrac{3a}{L}-4\left(\dfrac{a}{L}\right)^3\right]$
	$W(a/2+b/8)$	$\dfrac{W}{384\,EI}[8L^3-4Lb^2+b^3]$
	$Wa/3$	$\dfrac{Wa}{120\,EI}[16a^2-20ab+5b^2]$
	$WL/6$	$\dfrac{WL^3}{60\,EI}$
	$WL/8$	$\dfrac{WL^3}{73.14\,EI}$

Figure 4.15 Simply supported beam maximum moments and deflections

4.8.1 Bearing resistance of beam webs

The local bearing capacity of the web at its junction with the flange must be checked at supports and at points where loads are applied. The bearing capacity is given in Section 4.5.2 of BS 5950-1: 2000. An end bearing and an intermediate bearing are shown in Figures 4.16(a) and (b), respectively:

$$P_{bw} = (b_1 + nk)t p_{yw}$$

in which:

$n = 5$ for intermediate bearing,
$n = 2 + 0.6b_e/k$ but $n \le 5$ for end bearing,

and k is obtained as follows:

for a rolled I- or H-section: $k = T + r$,
for a welded I- or H-section: $k = T$,

(a)

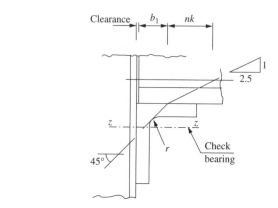

End bearing on a angle bracket

(b)

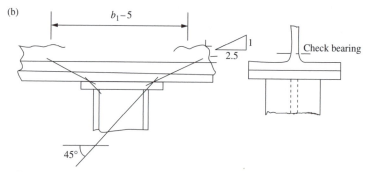

Intermediate bearing

Figure 4.16 Web and bracket bearing

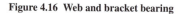

where

  b_1 is the stiff bearing length,
  b_e is the distance to the nearer end of the member from the end of the stiff
    bearing;
  p_{yw} is the design strength of the web;
  r is the root radius;
  T is the flange thickness:
  t is the web thickness.

The stiff bearing length b_1 is defined in Section 4.5.1.3 as the length which
cannot deform appreciably in bending. The dispersion of the load is taken as
45° through solid material. Stiff bearing lengths b_1 are shown in Figure 4.16(a).
For the unstiffened angle, a tangent is drawn at 45° to the fillet and the length
b_1 in terms of dimensions shown is:

$$b_1 = t + t + 0.8r - \text{clearance}$$

where r = radius of the fillet
    t = thickness of the angle leg

For the beam supported on the angle bracket as shown in Figure 4.16(a) the bracket is checked in bearing at Section ZZ and the weld or bolts to connect it to the column are designed for direct shear only. If the bearing capacity of the beam web is exceeded, stiffeners must be provided to carry the load (see Section 5.3.7).

4.8.2 Buckling resistance of beam webs

Types of buckling caused by a load applied to the top flange are shown in Figure 4.17. The web buckles at the centre if the flanges are restrained, otherwise sideways movement or rotation of one flange relative to the other occurs.

The buckling resistance of a web to loads applied through the flange is given in Section 4.5.3 of BS 5950-1: 2000. If the flange through which the load or reaction is applied is effectively restrained against both:

(a) rotation relative to the web; Figure 4.17(c) and
(b) lateral movement relative to the other flange; Figure 4.17(b),

then, provided that the distance α_e from the load or reaction to the nearer end of the member is at least $0.7d$, the buckling resistance of the unstiffened web should be taken as P_x given by:

$$P_x = \frac{25\varepsilon t}{\sqrt{(b_1 + nk)d}} P_{bw}$$

where d = depth of the web
P_{bw} = the bearing capacity of the unstiffened web at the web-to-flange connection.

If the distance α_e from the load or reaction to the nearer end of the member is less than $0.7d$, the buckling resistance P_x of the web should be taken as:

$$P_x = \frac{\alpha_e + 0.7d}{1.4d} \frac{25\varepsilon t}{\sqrt{(b_1 + nk)d}} P_{bw}$$

and b_1, k, n and t are as defined in Section 4.8.1. above.

This applies when the flange where the load is applied is effectively restrained against (a) rotation relative to the web and (b) lateral movement

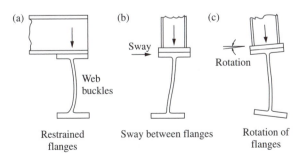

Figure 4.17 Types of buckling

relative to the other flange. Where (a) or (b) is not met, the buckling resistance of the web should be reduced to P_{xr}, given that:

$$P_{xr} = \frac{0.7d}{L_E}P_x$$

in which L_E is the effective length of the web acting as a compression member.

If the load exceeds the buckling resistance of the web, stiffeners should be provided (see Section 5.3.7).

4.8.3 Beam-end shear connections

Design procedures for flexible end shear connections for simply supported beams are set out here. The recommendations are from the SCI publication (11). Two types of shear connections, beam to column and beam to beam, are shown in Figures 4.18(a) and (b), respectively.

Design recommendations for the end plate are:

(1) Length—maximum = clear depth of web,
    minimum = 0.6 of the beam depth.
(2) Thickness—8 mm for beams up to 457 × 191 serial sizes,
    10 mm for larger beams.
(3) Positioning—the upper edge should be near the compression flange.

(a)
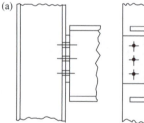
(b)
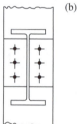
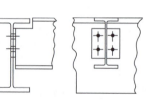

Beam to column joint   Beam to beam joint

(c)
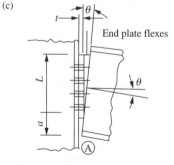
(d)
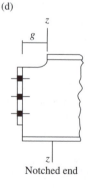

Rotation at beam support   Notched end

Figure 4.18 Flexible shear connection

Flexure of the end plate permits the beam end to rotate about its bottom edge, as shown in Figure 4.18(c). The end plate is arranged so that the beam flange at A does not bear on the column flange. The end rotation is taken as 0.03 radians, which represents the maximum slope likely to occur at the end of the beam. If the bottom flange just touches the column at A then

$$t/a = 0.03$$

or

a/t should be made < 33 to prevent contact.

The joint is subjected to shear only. The steps in the design are:

(1) Design the bolts for shear and bearing.
(2) Check the end plate in shear and bearing.
(3) Check for block shear.
(4) Design the weld between the end plate and beam web.

If the beam is notched as shown in Figure 4.18(d), the beam web should be checked for shear and bending at section z–z. To ensure that the web at the top of the notch does not buckle, the BCSA manual limits the maximum length of notch g to $24t$ for Grade S275 steel and $20t$ for Grade S355 steel, where t is the web thickness.

4.9 Examples of beam design

4.9.1 Floor beams for an office building

The steel beams for part of the floor of a library with book storage are shown in Figure 4.19(a). The floor is a reinforced concrete slab supported on universal beams. The design loading has been estimated as:

Dead load—slab, self weight of steel, finishes, ceiling,
$\qquad$ partitions, services and fire protection: $= 6.0\,\text{kN/m}^2$
Imposed load from Table I of BS 6399: Part 1 $\qquad = 4.0\,\text{kN/m}^2$

Determine the section required for beams 2A and 1B and design the end connections. Use Grade 275 steel.

The distribution of the floor loads to the two beams assuming two-way spanning slabs is shown in Figure 4.19:

(1) Beam 2A

Service dead load	$= 6 \times 3$	$= 18\,\text{kN/m}$,
Service imposed load	$= 4 \times 3$	$= 12\,\text{kN/m}$,
Factored shear	$= (1.4 \times 31.5) + (1.6 \times 21)$	$= 77.7\,\text{kN}$,
Factored moment	$= 1.4[(31.5 \times 2.5) - (13.5 \times 1.5)$	
	$\quad -(18 \times 0.5)] + 1.6[(21 \times 2.5)$	
	$\quad -(9 \times 1.5) - (12 \times 0.5)]$	$= 122.1\,\text{kN m}$.

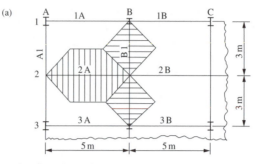

(a)

Part floor plan and load distribution

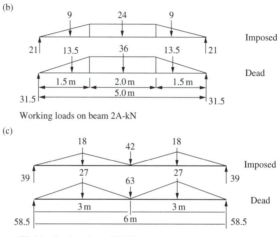

(b)

Working loads on beam 2A-kN

(c)

Working loads on beam B1-kN

Figure 4.19 Library: part floor plan and beam loads

Design strength, Grade 275—steel, thickness ≤ 16 mm, $p_y = 275$ N/mm^2,

$$\text{Plastic modulus } S = \frac{M}{p_y} = \frac{122.1 \times 10^3}{275} = 444 \text{ cm}^3,$$

Try 356×127 UB33 $S_x = 539.8$ cm^3, $Z_x = 470.6$ cm^3, $I_x = 8200$ cm^4.

The dimensions for the section are shown in Figure 4.20(a). The classification checks from Tables 11 BS 5950-1: 2000 are:

$$\varepsilon = (275/p_y)^{0.5} = 1.0$$
$$b/T = 62.7/8.5 = 7.37 < 9$$
$$d/T = 311.1/5.9 = 52.7 < 80$$

This is a plastic section.

The moment capacity is $p_y S \leq 1.2 p_y Z$

$$p_y S_x = 275 \times 539.8 \times 10^{-3} = 148.4 \text{ kN m},$$
$$1.2 p_y Z_x = 1.2 \times 275 \times 470.6 \times 10^{-3} = 155.3 \text{ kN m}.$$

The section is satisfactory for the moment.

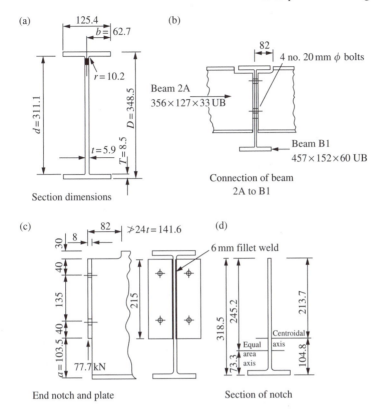

Figure 4.20 Section and end-connection beam 2A

The deflection due to the unfactored imposed load using formulae from Figure 4.15 is:

$$
\delta = \frac{18 \times 10^3 \times 1500}{120 \times 205 \times 10^3 \times 8200 \times 10^4}
$$
$$
\times [16 \times 1500^2 + 20 \times 1500 \times 2000 + 5 \times 2000^2]
$$
$$
+ \frac{24 \times 10^3}{384 \times 205 \times 10^3 \times 8200 \times 10^4}
$$
$$
\times [8 \times 5000^3 - 4 \times 5000 \times 2000^2 + 2000^3]
$$
$$
= 1.553 + 3.45
$$
$$
= 5.00 \, \text{mm}.
$$
$$
\delta/\text{span} = 5.00/5000 = 1/1000 < 1/360.
$$

The beam is satisfactory for deflection.

The end connection is shown in Figure 4.20(b) and the end shear is 77.7 kN. The notch required to clear the flange and fillet on beam B1 is shown in Figure 4.20(c). The end plate conforms to recommendations given in Section 4.8.3 above. To ensure end rotation:

$$a/t = 103.5/8 = 12.93 < 33$$

The bolts are 20 mm diameter, Grade 8.8:

Single shear value on threads $\qquad$ $= 91.9$ kN,
Capacity of four bolts $\qquad$ $= 4 \times 91.9$
$\qquad$ $= 367.6$ kN,
Bearing capacity of a bolt on 8 mm thick end plate $= 73.6$ kN,
Bearing on the end plate $\qquad$ $= 73.6$ kN.

Bolts and end plate are satisfactory in bearing. The web of beam B1 is checked for bearing below:

Shear capacity of end plate in shear on both sides

$$P_v = 2 \times 0.9 \times 0.6 \times 275 \times 8(215 - 44) \times 10^{-3} = 406.2 \text{ kN}$$

Provide 6 mm fillet weld in two lengths of 215 mm each.
The strength at 0.92 kN/mm $= 2(215 - 12) \times 0.92 = 374$ kN
Check the beam end in shear at the notch (see Figure 4.20(d))

$$P_v = 318 \times 5.9 \times 0.9 \times 0.6 \times 275 \times 10^{-3} = 279.0 \text{ kN}$$

Check the beam end in bending at the notch. The locations of the centroid and equal area axes of the T section are shown in Figure 4.20(d). The elastic and plastic properties may be calculated from first principles.
The properties are:
Elastic modulus top, $Z = 148.5 \text{ cm}^3$,
Plastic modulus, $S = 263.3 \text{ cm}^3$.

Moment capacity assuming a semi-compact section with the maximum stress limited to the design strength:

$$M_c = 148.5 \times 275 \times 10^{-3} = 40.8 \text{ kN m}$$

Factored moment at the end of the notch:

$$M = 77.7 \times 70 \times 10^{-3} = 5.44 \text{ kN m}$$

The beam end is satisfactory.
Note that the notch length 70 mm is taken from the *Guide to BS 5950-1: 2000, Vol.1, SCI.*

(2) Beam B1

The beam loads are shown in Figure 4.21(c). The point load at the centre is twice the reaction of Beam 2A. The triangular loads are:

Dead $= 2 \times 1.5^2 \times 6 = 27\,\text{kN}$,
Imposed $= 2 \times 1.5^2 \times 4 = 18\,\text{kN}$,
Factored shear $= (1.4 \times 58.5) + (1.6 \times 39) = 144.3\,\text{kN}$,
Factored moment $= 1.4[(58.5 \times 3) - (27 \times 1.5)] + 1.6[(39 \times 3)$
                     $- (18 \times 1.5)] = 333\,\text{kN m}$.
Plastic modulus, S $= 333 \times 10^3/275 = 1210.9\,\text{cm}^3$.
Try 457×152 UB 60: $S_x = 1284\,\text{cm}^3$, $Z_x = 1120\,\text{cm}^3$, $I_x = 25464\,\text{cm}^4$

The section is checked and found to be plastic.
    The moment capacity is $p_y S \le 1.2 p_y Z$

$$p_y S_x = 275 \times 1284 \times 10^{-3} = 353.1\,\text{kN m}$$
$$1.2 p_y Z_x = 1.2 \times 275 \times 1120 \times 10^{-3} = 369.6\,\text{kN m}$$

The section is satisfactory for the moment.

Shear capacity $P_v = 0.6 \times 275 \times 8.0 \times 454.7 \times 10^{-2}$
$\qquad\qquad\qquad = 600.2\,\text{kN (satisfactory)}$.

The deflection due to the unfactored imposed loads using formula from Figure 4.15 is:

$$\delta = \frac{42 \times 10^3 \times 6000^3}{48 \times 205 \times 10^3 \times 25464 \times 10^4} + \frac{36 \times 10^3 \times 6000^3}{73.14 \times 205 \times 10^3 \times 25464 \times 10^4}$$
$$= 5.65\,\text{mm}$$

$\delta/\text{span} = 5.65/6000 = 1/1062 < 1/360$ (satisfactory).
    The end connection is shown in Figure 4.21(a) with the beam supported on an angle bracket $150 \times 75 \times 10$ RSA. Details for the various checks are shown below:

(1) Bearing check (see Figure 4.21(c)):
    Bearing capacity, $P_{bw} = (b_1 + nk)t p_{yw} = (23.4 + 60.84) \times 8.0$
    $\qquad\qquad\qquad\qquad\qquad \times 275 \times 10^{-3} = 185.3\,\text{kN}$
    Satisfactory, Reaction $= 144.3\,\text{kN}$
(2) Buckling check (see Figure 4.21(b)):

$$P_x = \frac{\alpha_e + 0.7d}{1.4d}\,\frac{25\varepsilon t}{\sqrt{(b_1 + nk)d}}\,P_{bw}$$

$$P_x = \frac{23.4 + 0.7(407.7)}{1.4(407.7)}\,\frac{25(1.0)(8.0)}{\sqrt{(23.4 + 60.84)(407.7)}}(185)$$

$$= 107.5\,\text{kN}.$$

Web stiffener required, (see Chapter 5 for the design of stiffener)

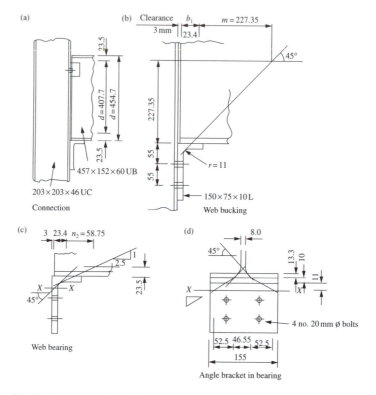

Figure 4.21 End-connection beam B1

(3) Check bracket angle for bearing at Section x-x (see Figures 4.21(d)):

Stiff bearing b_l $= 46.55\,\text{mm}$
Length in bearing $= 46.55 + 5 \times (10 + 11)$ $= 151.6\,\text{mm}$,
Bearing capacity $= 151.6 \times 10 \times 275 \times 10^{-3} = 417\,\text{kN}$.

(4) Bracket bolts:
   Provide four No. 20 mm diameter Grade 8.8 bolts.
   Shear capacity $= 4 \times 91.9 = 367.6\,\text{kN}$.
   Satisfactory, the bolts are adequate.

(5) Check beam B1 for bolts from 2 No. beams 2A bearing on web:
   For 8.0 mm thick:

Reactions $= 2 \times 77.7$ $= 155.4\,\text{kN}$,
Bearing capacity of bolts $= 4 \times 460 \times 20 \times 8.0 \times 10^{-3} = 294.4\,\text{kN}$.

The joint is satisfactory.

4.9.2 Beam with unrestrained compression flange

Design the simply supported beam for the loading shown in Figure 4.22. The loads P are normal loads. The beam ends are restrained against torsion with the compression flange free to rotate in plan. The compression flange is unrestrained between supports. Use Grade S275 steel.

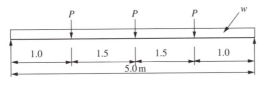

$P = 25\,\text{kN dead load}$
$12\,\text{kN imposed load}$
$W = 2.0\,\text{kN/m dead load}$

Figure 4.22 Beam with unrestrained compression flange

Factored load $\quad = (1.4 \times 37.5) + (1.4 \times 5) + (1.6 \times 18) = 88.3\,\text{kN},$
Factored moment $= 1.4(37.5 \times 2.5 - 25 \times 1.5) + 1.4 \times 2 \times 5^2/8$
$\quad\quad\quad\quad\quad\quad +1.6(18 \times 2.5 - 12 \times 1.5) = 130.7\,\text{kN m}.$

Try 457×152 UB 60. The properties are:

$$r_y = 3.23\,\text{cm}; \quad x = 37.5, \quad u = 0.869, \quad S_x = 1280\,\text{cm}^3.$$

Note that a check will confirm this is a plastic section.
   Design strength $p_y = 275\,\text{N/mm}^2$ (Table 9, BS 5950)
   The effective length, L_E from Table 13 of BS 5950-1: 2000:

$$L_E = 1.0 L_{LT} = 5000\,\text{mm}.$$

Equivalent slenderness $\lambda_{LT} = u v \lambda \sqrt{\beta_w}$

$\lambda = 5000/32.3 \quad = 154.8,$
$\eta = 0.5 \quad \text{and} \quad x = 37.5,$
$\lambda/x = 154.8/37.5 = 4.13.$

$v = 0.855$ from Table 19 of BS 5950-1: 2000,

$$\lambda_{LT} = 0.869 \times 0.855 \times 154.8 \times 1.0 = 115.$$

Bending strength, $p_b = 102\,\text{N/mm}^2$ (Table 17, BS 5950)
Buckling resistance moment:

$M_b = 102 \times 1280 \times 103 = 130.6\,\text{kN m}$
$m_{LT} = 0.925$
$M_b/m_{LT} = 130.6/0.925 = 141.2\,\text{kN m}$

Shear capacity:

   Overall depth, $D = 454.7\,\text{mm},$
   Web thickness, $t = 8.0\,\text{mm},$

$$P_v = 0.6 \times 275 \times 454.7 \times 8 \times 10^{-3} = 600.2\,\text{kN},$$

The section is satisfactory.

The conservative approach in Section 4.3.7. of BS 5950-1: 2000 gives:

$L_E/r_y = 154.8$ $\sqrt{\beta_w} = 1.0$ $D/T = 454.7/13.3 = 34.2$,

$p_b = 100.0 \, \text{N/mm}^2$—(Table 20, BS 5950),

$M_b = 100.0 \times 1280 \times 10^{-3} = 128.0 \, \text{kN m}$,

$M_b/m_{LT} = 128.0/0.925 = 138.4 \, \text{kN m}$.

The section is satisfactory.

4.9.3 Beam subjected to bending about two axes

A beam of span 5 m with simply supported ends not restrained against torsion has its major principal axis inclined at 30° to the horizontal, as shown in Figure 4.23. The beam is supported at its ends on sloping roof girders. The unrestrained length of the compression flange is 5 m. If the beam is 457 × 152 UB 52, find the maximum factored load that can be carried at the centre. The load is applied by slings to the top flange.

Let the centre factored load $= W$ kN. The beam self weight is unfactored.

$$\text{Moments } M_x = [W \times 5/4 + (1.4 \times 52 \times 9.81 \times 5^2 \times 10^{-3})/8]\cos 30°$$
$$= 1.083 \, W + 1.933$$
$$M_y = M_x \tan 30° = 0.625 \, W + 1.116.$$

Properties for 457 × 152 UB 52:

$S_x = 1090 \, \text{cm}^3$, $Z_y = 84.6 \, \text{cm}^3$, $r_y = 3.11 \, \text{cm}$,
$x = 43.9$, $u = 0.859$.

The section is a plastic section. The design strength $p_y = 275 \, \text{N/mm}^2$ (Table 9, BS 5950).

(1) Moment capacity for x–x axis

Effective length: the ends are torsionally unrestrained and free to rotate in plan and the load is destabilizing. (Refer to Table 13 of BS 5950.)

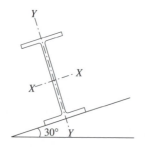
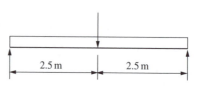

Figure 4.23 Beam in biaxial bending

$L_E = 1.4L_{LT} + 2D,$
$L_E = 1.4(5000) + 2 \times 449.8 = 7899.6$ mm.
Slenderness $\lambda = 7899.6/31.1 = 254.0$.
The load is destabilizing, $m_{LT} = 1.0$ (Clause 4.3.6.6 BS 5950).

$\eta = 0.5$, uniform I section.
$\lambda/x = 254.0/43.9 = 5.79.$
$v = 0.778$ (Table 19 of BS 5950).

Equivalent slenderness:

$$\lambda_{LT} = 0.859 \times 0.768 \times 254 \times 1.0 = 167.6.$$

Bending strength, $p_b = 55.5$ N/mm^2 (Table 16 of BS 5950).
Buckling resistance moment, $M_b = 55.5 \times 1090 \times 10^{-3} = 60.5$ kN m.

(2) Biaxial bending

The capacity in biaxial bending is determined by the buckling capacity at the centre of the beam (see Section 4.5.4). The interaction relationships to be satisfied are:

$$\frac{m_x M_x}{p_y Z_x} + \frac{m_y M_y}{p_y Z_y} \leq 1$$

$$\frac{m_{LT} M_{LT}}{M_b} + \frac{m_y M_y}{p_y Z_y} \leq 1$$

The moment capacity for the x–x axis:

$$p_y Z_x = 275 \times 950 \times 10^{-3} = 261.3 \text{ kN m.}$$

The moment capacity for the y–y axis:

$$p_y Z_y = 275 \times 84.6 \times 10^{-3} = 23.3 \text{ kN m.}$$

Factor m_x and $m_y = 0.9$ (Table 26 BS 5950)

$$\frac{0.9(1.083W + 1.933)}{261.3} + \frac{0.9(0.625W + 1.116)}{23.3} = 1, \qquad (4.1)$$

$W = 34.1$ kN

$$\frac{1.0(1.083W + 1.933)}{60.5} + \frac{0.9(0.625W + 1.116)}{23.3} = 1. \qquad (4.2)$$

$W = 22.0$ kN

Clearly, Equation (2) is more critical, therefore the maximum that the beam can be carried at the centre is 22.0 kN.

4.10 Compound beams

4.10.1 Design considerations

A compound beam consisting of two equal flange plates welded to a universal beam is shown in Figure 4.24.

(1) Section classification

Compound sections are classified into plastic, compact, semi-compact and slender in the same way as discussed for universal beams in Section 4.3. However, the compound beam is treated as a section built up by welding. The limiting proportions from Table 11 of BS 5950-1: 2000 for such sections are shown in Figure 6 of the code. The manner in which the checks are to be applied set out in Section 3.5.3 of the code is as follows:

(1) Whole flange consisting of flange plate and universal beam flange is checked using b_1/T, where b_1 is the total outstand of the compound beam flange and T the thickness of the original universal beam flange.
(2) The outstand b_2 of the flange plate from the universal beam flange is checked using b_2/T_f, where T_f is the thickness of the flange plate.
(3) The width/thickness ratio of the flange plate between welds b_3/T_f is checked, where b_3 is the internal width of the universal beam flange.
(4) The universal beam flange itself and the web must also be checked as set out in Section 4.3.

(2) Moment capacity

The area of flange plates to be added to a given universal beam to increase the strength by a required amount may be determined as follows. This applies to a restrained beam (see Figure 4.25(a)):

Total plastic modulus required:

$$S_x = M/p_y$$

where M is the applied factored moment.

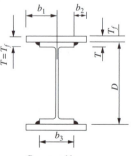

Compound beam

Figure 4.24 Compound beams fabricated by welding

(a)

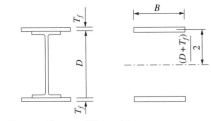

Compound beam and flange plates

(b)

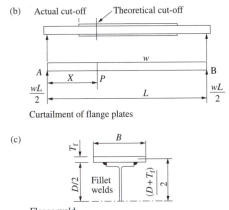

Curtailment of flange plates

(c)

Flange weld

Figure 4.25 Compound beam design

If S_{UB} is the plastic modulus for the universal beam, the additional plastic modulus required is:

$$S_{ax} = S_x - S_{UB} = 2BT_f(D + T_f)/2$$

where $B \times T_f$ is the flange area and D is the depth of universal beam.

Suitable dimensions for the flange plates can be quickly established. If the beam is unrestrained, successive trials will be required.

(3) Curtailment of flange plates

For a restrained beam with a uniform load the theoretical cut-off points for the flange plates can be determined as follows (see Figure 4.25(b)):

The moment capacity of the universal beam:

$$M_{UB} = p_y S_{UB} \leq 1.2 p_y Z_{UB}$$

where Z_{UB} is the elastic modulus for the universal beam.

Equate M_{UB} to the moment at P a distance x from the support:

$$wLx/2 - wx^2/2 = M_{UB}$$

where w is the factored uniform load and L the span of the beam.

Solve the equation for x. The flange plate should be carried beyond the theoretical cut-off point so that the weld on the extension can develop the load in the plate at the theoretical cut-off.

(4) Web

The universal beam web must be checked for shear. It must also be checked for buckling and crushing if the beam is supported on a bracket or column or if a point load is applied to the top flange.

(5) Flange plates to universal beam welds

The fillet welds between the flange plates and universal beam are designed to resist horizontal shear using elastic theory (see Figure 4.25(c)).
    Horizontal shear in each fillet weld:

$$\frac{F_v B T_f (D - T_f)}{4 I_x}$$

where F_v is the factored shear and I_x is the moment of inertia about x–x axis. The other terms have been defined above.
    The leg length can be selected from Table 10.5. In some cases a very small fillet weld is required, but the minimum recommended size of 6 mm should be used.
    Intermittent welds may be specified, but continuous welds made automatically are to be preferred. These welds considerably reduce the likelihood of failure due to fatigue or brittle fracture.

4.10.2 Design of a compound beam

A compound beam is to carry a uniformly distributed dead load of 400 kN and an imposed load of 600 kN. The beam is simply supported and has a span of 11 m. Allow 30 kN for the weight of the beam. The overall depth must not exceed 700 mm. The length of stiff bearing at the ends is 215.9 mm where the beam is supported on 203 × 203 UC 71 columns. Full lateral support is provided for the compression flange. Use Grade S275 steel.

(1) Design the beam section and check deflection, assuming a uniform section throughout.
(2) Determine the theoretical and actual cut-off points for the flange plates and the possible saving in weight that would result if the flange plates were curtailed.
(3) Check the web for shear, buckling and bearing, assuming that plates are not curtailed.
(4) Design the flange plate to universal beam welds.

(1) Design of the beam section

The total factored load carried by the beam $= 1.4(400 + 30) + (1.6 \times 600) = 1562$ kN (i.e. 142 kN/m).
    Maximum moment $= 1562 \times 11/8 = 2147.8$ kN m.

(a)

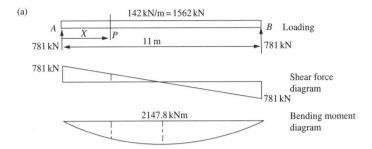

Loading, shear force and bending moment diagrams

(b)

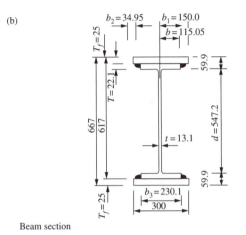

Beam section

Figure 4.26 Compound beam

The loading, shear force and bending moment diagrams are shown in Figure 4.26(a).

Assume that the flanges of the universal beam are thicker than 16 mm:

$$p_y = 265 \, \text{N/mm}^2 \text{ (from Table 9, BS 5950)}$$

Plastic modulus required, $S_x = 2147.8 \times 10^3/265 = 8104.9 \, \text{cm}^3$.
Try 610×229 UB 140, where $S_x = 4146 \, \text{cm}^3$.

The beam section is shown in Figure 4.26(b).
The additional plastic modulus required:

$$= 8104.9 - 4146 = 3958.9 \, \text{cm}^3$$
$$= 2 \times 300 \times T_f(617 + T_f)/(2 \times 10^3),$$

where the flange plate thickness T_f is to be determined for a width of 300 mm. This reduces to:

$$T_f^2 + 617T_f - 13196 = 0.$$

Solving gives $T_f = 20.69$ mm.
Provide plates 300 mm × 25 mm.
The total depth is 667 mm (satisfactory).
Check the beam dimensions for local buckling:

$$\varepsilon = (275/265)^{0.5} = 1.02.$$

Universal beam (see Figure 4.4):

Flange: $b/T = 115.1/22.1 = 5.21 < 9.0 \times 1.02 = 9.18$,
Web: $d/t = 547.2/13.1 = 41.7 < 80 \times 1.02 = 81.6$.

Compound beam flange (see Figure 4.25):

Flange $b_1/T = 150/22.1 = 6.79 < 1.02 \times 8.0 = 8.16$
$b_2/T_f = 34.95/25 = 1.40 < 8.16$
$b_3/T_f = 230.1/25 = 9.2 < 28 \times 1.02 = 28.56$.

The section meets the requirements for a plastic section.
    The moment of inertia about the $x-x$ axis for the compound section is
calculated. Note for the universal beam:

$$I_x = 111844 \, \text{cm}^3,$$

$$I_x = 111844 + 2 \times 30 \times 2.5 \times 32.1^2 + 2 \times 30 \times 2.5^2/12$$

$$= 266483 \, \text{cm}^4.$$

The deflection due to the unfactored imposed load is

$$\delta = \frac{5 \times 600 \times 10^3 \times 11000^3}{384 \times 205 \times 10^3 \times 266483 \times 10^4} = 19.03 \, \text{mm}$$

$\delta/\text{span} = 19.03/11000 = 1/578 < 1/360$ (Satisfactory)

(2) Curtailment of flange plates

Moment capacity of the universal beam:

$$M_c = 4146 \times 265 \times 10^{-3} = 1098.7 \, \text{kN m}.$$

Referring to Figure 4.26(a), determine the position of P where the bending
moment in the beam is 1098.7 kN m from the following equation:

$$781x - 142x^2/2 = 1098.7$$

This reduces to

$$x^2 - 11x + 15.47 = 0$$

$x = 1.656$ m from each end.

The compound section will be the elastic range at this point with an average stress in the plate for the factored loads

$$\frac{1098.7 \times 10^6 \times 321}{266483 \times 10^4} = 132.4 \, \text{N/mm}^2,$$

Force in the flange plate

$$= 132.4 \times 300 \times 25 \times 10^{-3} = 993 \, \text{kN}.$$

Assume 6 mm fillet weld, strength 0.92 kN/mm from Table 4.5 (see (4) below). Length of weld to develop the force in the plate

$$= [993/(2 \times 0.92)] + 6 = 546 \, \text{mm},$$

Actual cut-off length $= 1656 - 546 = 1110 \, \text{mm}.$
Cut plates off at 1000 mm from each end.
    Saving in material from curtailment:

Area of universal beam $= 178.4 \, \text{cm}^2,$
Area of flange plates $= 150 \, \text{cm}^2.$

Volume of the compound beam with no curtailment of plates

$$= 328.4 \times 1100 = 36.12 \times 10^4 \, \text{cm}^3.$$

Volume of material saved $= 200 \times 150 = 3.0 \times 10^4 \, \text{cm}^3.$
Saving in material $= 8.3\%$

(3) Web in shear, buckling and bearing

(1) Shear capacity (see Figure 4.26(b)). This is checked on the web of the universal beam.

$$P_v = 0.6 \times 265 \times 617 \times 13.1 \times 10^{-3} = 1285 \, \text{kN},$$
Factored shear, $F_v = 781 \, \text{kN}.$

(2) Web bearing (see Figures 4.26(b) and 4.27(a)):

$$P_{bw} = (b_1 + nk)t p_{yw}$$
$$P_{bw} = (215.9 + 149.75) \times 13.1 \times 265 \times 10^{-3}$$
$$= 1269.3 \, \text{kN (satisfactory)}.$$

(3) Web buckling:

$$P_x = \frac{\alpha_e + 0.7d}{1.4d} \frac{25\varepsilon t}{\sqrt{(b_1 + nk)d}} P_{bw}$$
$$P_x = \frac{108 + 0.7 \times (547.2)}{1.4 \times (547.2)} \frac{25 \times 1.0 \times 13.1}{\sqrt{(215.9 + 149.75)547.2}} 1269.3$$
$$= 595 \, \text{kN}.$$

Web stiffeners required. (see Chapter 5 for the design of stiffener)

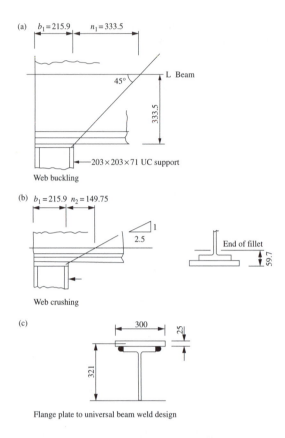

(a) $b_1 = 215.9$ $n_1 = 333.5$

45°

L Beam

333.5

203×203×71 UC support

Web buckling

(b) $b_1 = 215.9$ $n_2 = 149.75$

1

2.5

End of fillet

59.7

(c)

300

25

321

Flange plate to universal beam weld design

Figure 4.27 Bearing and buckling check and flange weld design

(4) Flange plate to universal beam weld (see Figure 4.27(b)):
   Factored shear at support $F_v = 781$ kN

$$\text{Horizontal shear on two fillet welds} = \frac{781 \times 300 \times 25 \times 321}{2 \times 266483 \times 10^4}$$
$$= 0.353 \text{ kN/mm}.$$

Provide 6 mm fillet welds, strength 0.92 kN/mm. This is the minimum size weld to be used.

4.11 Crane beams

4.11.1 Types and uses

Crane beams carry hand-operated or electric overhead cranes in industrial buildings such as factories, workshops, steelworks, etc. Types of beams used are shown in Figures 4.28(a) and (b). These beams are subjected to vertical and horizontal loads due to the weight of the crane, the hook load and dynamic loads. Because the beams are subjected to horizontal loading, a larger flange or horizontal beam is provided at the top on all but beams for very light cranes.

   Light crane beams consist of a universal beam only or of a universal beam and channel, as shown in Figure 4.28(a). Heavy cranes require a plate girder

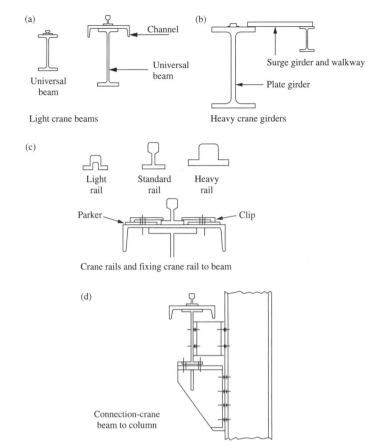

Figure 4.28 Type of crane beams and rails and connection to column

with surge girder, as shown in Figure 4.28(b). Only light crane beams are considered in this book. Some typical crane rails and the fixing of a rail to the top flange are shown in Figure 4.28(c). The connection of a crane girder to the bracket and column is shown in Figure 4.28(d). The size of crane rails depends on the capacity and use of the crane.

4.11.2 Crane data

Crane data can be obtained from the manufacturer's literature. The data required for crane beam design are:

Crane capacity
Span
Weight of crane
Weight of the crab
End carriage wheel centres
Minimum hook approach
Maximum static wheel load

The data are shown in Figure 4.29.

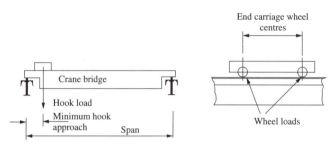

Figure 4.29 Crane design data

(1) Loads on crane beams

Crane beams are subjected to:

(1) Vertical loads from self weight, the weight of the crane, the hook load and impact; and
(2) horizontal loads from crane surge.

Cranes are classified into four classes in BS 2573: *Rules for Design of Cranes*, Part 1: Specification for Classification, Stress Calculations and Design Criteria for Structures. The classes are:

Class 1—light. The safe working load is rarely hoisted;
Class 2—moderate. The safe working load is hoisted fairly frequently;
Classes 3 and 4 are heavy and very heavy cranes.

Only beams for cranes of classes 1 and 2 are considered in this book. The dynamic loads caused by these classes of cranes are given in BS 6399: Part 1, Section 7. The loading specified in the code is set out below.

The following allowances shall be deemed to cover all forces set up by vibration, shock from slipping of slings, kinetic action of acceleration and retardation and impact of wheel loads:

(1) For loads acting vertically, the maximum static wheel loads shall be increased by the following percentages:
For electric overhead cranes: 25%
For hand-operated cranes: 10%
(2) The horizontal force acting transverse to the rails shall be taken as a percentage of the combined weight of the crab and the load lifted as follows:
For electric overhead cranes: 10%
For hand-operated cranes: 5%
(3) The horizontal force acting along the rails shall be taken as 5 per cent of the static wheel loads for either electric or hand-operated cranes.

The forces specified in (2) or (3) may be considered as acting at rail level. Either of these forces may act at the same time as the vertical load. The load factors to be used with crane loads given in Table 2 in the code are:

Vertical or horizontal crane loads considered separately:

$$\gamma_f = 1.6$$

Vertical and horizontal crane loads acting together:

$$\gamma_f = 1.4$$

The application of these clauses will be shown in an example.

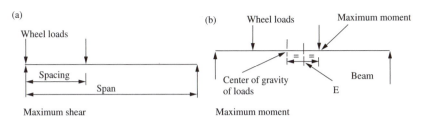

Figure 4.30 Rolling loads: maximum shear and moment

(2) Maximum shear and moment

The wheel loads are rolling loads, and must be placed in position to give maximum shear and moment. For two equal wheel loads:

(1) The maximum shear occurs when one load is nearly over a support;
(2) The maximum moment occurs when the centre of gravity of the loads and one load are placed equidistant about the centre line of the girder. The maximum moment occurs under the wheel load nearest the centre of the girder.

The load cases are shown in Figure 4.30.

Note that if the spacing between the loads is greater than 0.586 of the span of the beam, the maximum moment will be given by placing one wheel load at the centre of the beam.

4.11.3 Crane beam design

(1) Buckling resistance moment for x–x axis

Section properties

A crane girder section consisting of a universal beam and channel is shown in Figure 4.31(a) and the elastic properties for a range of sections are given in the *Structural Steelwork Handbook*. The plastic properties are calculated as follows for the plastic stress distribution with bending about the equal area axis shown in Figure 4.31(b).

First locate the equal area axis by trial and error and then calculate the positions of the centroids of the tension and compression areas. If z is the lever arm between these centroids, the plastic modulus:

$$S_x = Az/2$$

where A is the total area of cross-section.

The plastic modulus may also be calculated from the definition. This is the algebraic sum of the first moments of area about the equal area axis.

Lateral torsional buckling

The code specifies in Section 4.11.3 that no reduction is to be made for the equivalent uniform moment factor $m_{LT} = 1.0$. The effective length $L_E =$ span for a simply supported beam with the ends torsionally restrained and the compression flange laterally restrained but free to rotate on plan.

The slenderness $\lambda = L_E/r_y$, where $r_y = $ radius of gyration for the whole section about the y–y axis.

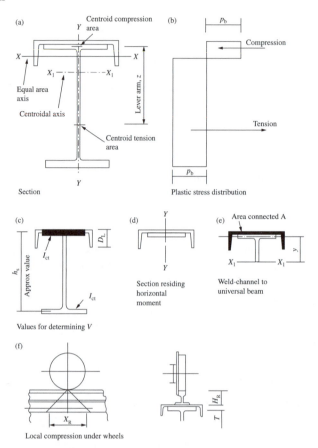

Figure 4.31 Column beam design

The factors modifying the slenderness are set out in Appendix B.2.4 of the code. The buckling parameter $u = 1.0$. This may also be calculated from a formula in Appendix B. The torsional index x for a flanged section symmetrical about the minor axis is:

$$x = 0.566\,h_s\,(A/J)^{0.5}$$

where h_s is the distance between the shear centres of the flanges.

As a conservative approximation, h_s may be taken as the distance from the centre of the bottom flange to the centroid of the channel web and universal beam flange, as shown in Figure 4.31(c):

A = area of cross-section,
J = torsion constant = $1/3(\Sigma bt^3 + h_w t_w^3)$,
b = flange width,
t = flange thickness,
h_w = web depth,
t_w = web thickness.

Note that the top flange of the universal beam and channel web act together, so t is the sum of the thicknesses. The width may be taken as the average of the widths of the universal beam flange and the depth of web of the channel:

$$\eta = \frac{I_{cf}}{I_{cf} + I_{tf}}$$

where I_{cf} is the moment of inertia of the top flange about the y–y axis $= I_x$ (channel) $+(1/2)\, I_y$ (universal beam),
I_{tf} is the moment of inertia of the bottom flange about the y–y axis $= (1/2)\, I_y$ (universal beam).

The monosymmetry index ψ for an I- or T-section with lipped flange is:

$$\psi = 0.8(2\eta - 1)(1 + 0.5D_L/D)$$

where D denote overall depth of section, D_L the depth of lip and the breadth of channel flange.

The slenderness factor:

$$v = \frac{1}{[(4\eta(1 - \eta) + 0.05(\lambda/x)^2 + \psi^2)^{0.5} + \psi]^{0.5}}$$

The modified slenderness:

$$\lambda_{LT} = uv \cdot \lambda \cdot \sqrt{\beta_w}$$

The bending strength p_b is obtained from Table 17 for welded sections.
The buckling resistance moment:

$$M_b = S_x p_b$$

This must exceed the factored moment for the vertical loads only including impact with load factor 1.6.

(2) Moment capacity for the y–y axis (see Figure 4.31(d))

The horizontal bending moment is assumed to be taken by the channel and top flange of the universal beam. The elastic modulus Z_y for this section is given in the *Structural Steelwork Handbook*. The moment capacity:

$$M_{cy} = Z_y p_y$$

(3) Biaxial bending check

The overall buckling check using the simplified approach is given in Section 4.8.3 of BS 5950-1: 2000. This is:

$$\frac{M_x}{p_y Z_x} + \frac{M_y}{p_y Z_y} \leq 1$$

$$\frac{M_x}{M_b} + \frac{M_y}{p_y Z_y} \leq 1$$

Two checks are required:

(1) Vertical crane loads with no impact and horizontal loads only with load factor 1.6; and
(2) Vertical crane loads with impact and horizontal loads both with load factor 1.4.

(4) Shear capacity

The vertical shear capacity is checked as for a normal beam (see Section 4.6.2). The horizontal shear load is small and is usually not checked.

(5) Weld between channel and universal beam (see Figure 4.31(e))

The horizontal shear force in each weld:

$$\frac{FAy}{2I_x}$$

where F = factored shear,
A = area connected by the weld = area of the channel,
y = distance from the centroid of the channel to the centroid of the crane beam
I_x = moment of inertia of the crane beam about the x–x axis.

The elastic properties are given in the *Structural Steelwork Handbook*.

(6) Web buckling and bearing

The web is to be checked for buckling and bearing as set out in Section 4.8. The length to be taken for stiff bearing depends on the bracket construction or other support for the crane beam (for example, if it is carried on a crane column).

(7) Local compression under wheels (see Figure 4.31(f))

BS 5950-1: 2000 specifies in Section 4.11.4 that the local compression on the web may be obtained by distributing the crane wheel load over a length:

$$x_R = 2(H_R + T) \quad \text{but} \quad x_R \le s_w$$

where

H_R = rail height;
S_w = the minimum distance between centres of adjacent wheels;
T = flange thickness.

Bearing stress = $p/(tx_R)$

where

p = crane wheel load
t = web thickness.

This stress should not exceed the design strength of the web p_{yw}.

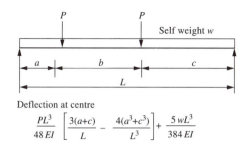

Deflection at centre

$$\frac{PL^3}{48\,EI}\left[\frac{3(a+c)}{L} - \frac{4(a^3+c^3)}{L^3}\right] + \frac{5\,wL^3}{384\,EI}$$

Figure 4.32 Crane beam deflection

4.11.4 Crane beam deflection

The deflection limitations for crane beams given in Table 8 of BS 5950-1: 2000 are quoted in Table 2.2 in this book. These are:

(1) Vertical deflection due to static wheel loads = span/600.
(2) Horizontal deflection due to crane surge, calculated using the top flange properties alone = span/500.

The formula for deflection at the centre of the beam is given in Figure 4.32 for crane wheel loads placed in the position to give the maximum moment. The deflection should also be checked with the loads placed equidistant about the centre of the beam, when $a = c$ in the formula given.

4.11.5 Design of a crane beam

Design a simply supported beam to carry an electric overhead crane. The design data are as follows:

Crane capacity	= 100 kN,
Span between crane rails	= 20 m,
Weight of crane	= 90 kN,
Weight of crab	= 20 kN,
Minimum hook approach	= 1.1 m,
End carriage wheel centres	= 2.5 m,
Span of crane girder	= 5.5 m,
Self weight of crane girder	= 8 kN.

Use Grade S275 steel.

(1) Maximum wheel loads, moments and shear

The crane loads are shown in Figure 4.33(a). The maximum static wheel loads at A

$$= \frac{90}{4} + \frac{120 \times 18.9}{20 \times 2} = 79.2\,\text{kN}.$$

The vertical wheel load, including impact

$$= 79.2 + 25\% = 99\,\text{kN}.$$

The horizontal surge load transmitted by friction to the rail through four wheels:

$$= 10\%(100 + 20)/4 = 3\,\text{kN}.$$

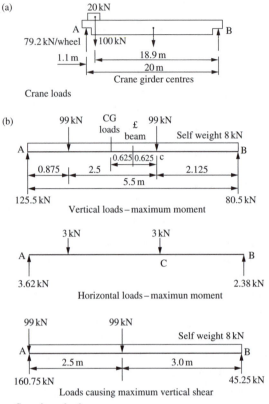

(a)

Crane loads

(b)

Vertical loads – maximum moment

Horizontal loads – maximun moment

Loads causing maximum vertical shear

Crane beam loads

Figure 4.33 Crane and crane beam loads

Load factors from Table 2.1:

Dead load − self weight
$$\gamma_f = 1.4$$

Vertical and horizontal crane loads considered separately

$$\gamma_f = 1.6,$$

Vertical and horizontal crane loads acting together

$$\gamma_f = 1.4.$$

The crane loads in a position to give maximum vertical and horizontal moments and maximum vertical shear are shown in Figure 4.33(b). The maximum vertical moments due to dead load and crane loads are calculated separately:
   Dead load

$$R_B = 4\,\text{kN},$$
$$M_c = (4 \times 2.125) - (8 \times 2.125^2/(5.5 \times 2)) = 5.22\,\text{kN m}.$$

Crane load, including impact

$$R_B = 99(0.875 + 3.375)/5.5 = 76.5\,\text{kN},$$
$$M_c = 76.5 \times 2.125 = 162.6\,\text{kN m}.$$

Crane loads with no impact

$$M_c = 162.6 \times 79.2/99 = 130.1\,\text{kN m}.$$

The maximum horizontal moment due to crane surge

$$R_B = 3(0.875 + 3.375)/5.5 = 2.32\,\text{kN},$$
$$M_c = 2.32 \times 2.125 = 4.93\,\text{kN m}.$$

The maximum vertical shear

Dead load $R_A = 4\,\text{kN}.$

Crane loads, including impact

$$R_A = 99 + 99 \times 3.0/5.5 = 153.0\,\text{kN}.$$

The load factors are introduced to calculate the design moments and shear for the various load combinations:

(1) Vertical crane loads with impact and no horizontal crane load.
    Maximum moment
    $M_c = (1.4 \times 5.22) + (1.6 \times 162.6) = 267.5\,\text{kN m},$
    Maximum shear
    $F_A = (1.4 \times 4) + (1.6 \times 153.0) = 250.4\,\text{kN}.$
(2) Horizontal crane loads and vertical crane loads with no impact.
    Maximum horizontal moment
    $M_c = 1.6 \times 4.93 = 7.89\,\text{kN m},$
    Maximum vertical moment
    $M_c = (1.4 \times 5.22) + (1.6 \times 130.1) = 215.47\,\text{kN m}.$
(3) Vertical crane loads with impact and horizontal crane loads acting together.
    Maximum vertical moment
    $M_c = (1.4 \times 5.22) + (1.4 \times 162.6) = 234.95\,\text{kN m},$
    Maximum horizontal moment
    $M_c = 1.4 \times 4.93 = 6.9\,\text{kN m}.$

(2) Buckling resistance moment for the x–x axis

The following trial section is selected:

457×191 UB $74 + 254 \times 76$ Channel.

Referring to Figure 4.34, the equal area axis x–x and centroids of the tension and compression areas are located and the plastic modulus calculated. Computations are shown in the figure. The elastic properties for this crane beam are taken from the *Structural Steelwork Handbook,* and these are shown in Figure 4.35.

(a)

(b)

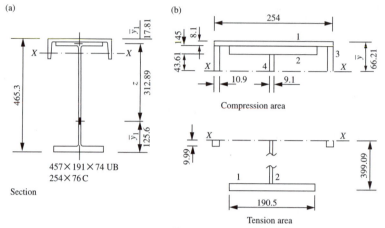

Section

457×191×74 UB
254×76 C

Compression area

Simplified section

Tension area

Figure 4.34 Crane beam: plastic properties

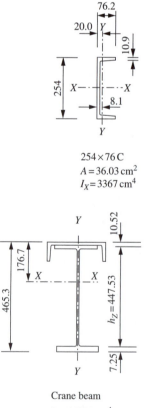

254×76 C
$A = 36.03 \, \text{cm}^2$
$I_X = 3367 \, \text{cm}^4$

457×191×74 UB
$I_Y = 1671 \, \text{cm}^4$

Crane beam
$I_X = 45983 \, \text{cm}^4$
$I_Y = 6.2 \, \text{cm}$

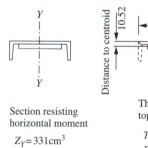

Section resisting
horizontal moment
$Z_Y = 331 \, \text{cm}^3$
$I_Y = 4202 \, \text{cm}^4$

Thickness of
top flange
$T = A/b$
$D/T = 24.5$

Figure 4.35 Crane beam: elastic properties

1. Locate equal area axis

$$\text{Total area} = (25.4 \times 0.81) + (6.81 \times 2 \times 1.09) + (2 \times 19.05 \times 1.45)$$
$$+ (42.82 \times 0.91) = 129.64 \, \text{cm}^2.$$
$$64.82 = (25.4 \times 0.82) + (19.05 \times 1.45) + 2 \times 1.09(y - 0.81)$$
$$+ 0.91(y - 1.45 - 0.81)$$
$$\bar{y} = 6.618 \, \text{cm}$$

2. Locate centroids of compression and tension areas

No	Area	y	Ay	No	Area	y	Ay
\multicolumn{4}{c}{*Compression area* *area moments about top*}				\multicolumn{4}{c}{*Tension area* *area moments about bottom*}			
1	20.54	0.405	8.32	1	27.62	0.725	20.02
2	27.62	1.535	42.39	2	35.00	20.27	709.5
3	12.67	3.175	47.07	3	2.18	38.59	84.12
4	3.97	4.441	17.63				
Sum	64.83		115.41		64.80		813.59
	$y_1 = 1.78$				$y_2 = 12.56$		

3. Lever arm, $Z = 46.53 - 1.78 - 12.56 = 31.19 \, \text{cm}$
4. Plastic modulus, $S_x = 64.83 \times 32.19 = 2086.8 \, \text{cm}^3$

The bending strength, p_b taking lateral torsional buckling into account, is determined:

Effective length $L_E = \text{span} = 5500$,
Slenderness $\lambda = L_E/r_y = 5500/62 = 88.7$,
Factors modifying slenderness:
Buckling parameter, $u = 1.0$.

This is conservative: the value of 0.81 is calculated from the formula in Appendix B of the code.
The slenderness factor v is calculated from the formulae in Appendix B:

$$I_{cf} = I_x(\text{channel}) + \tfrac{1}{2}I_y(\text{UB})$$
$$= 3367 + 835.5 = 4202.5 \, \text{cm}^4,$$
$$I_{tf} = \tfrac{1}{2}I_y(\text{UB}) = 835.5 \, \text{cm}^4.$$
$$\eta = \frac{4202.5}{4202.5 + 835.5} = 0.834.$$

The distance between the shear centres of the flanges:

h_s = distance from centre of bottom flange to centroid of channel web and universal beam flange = 447.53 mm approximately (see Figure 4.35).
Torsion constant:

$$J = \tfrac{1}{3}[(14.5^2 \times 190.5) + (9.1^2 \times 428.2)$$
$$+ (22.6^2 \times 211.35) + (2 \times 10.93 \times 76.2)]$$
$$= 1.18 \times 10^6 \text{ mm}^4.$$
$$\text{Area} A = 12964 \text{ mm}^2.$$

The torsional index

$$x = 0.566 \times 447.53(13103/1.18 \times 10^6)^{0.5} = 26.5.$$

This compares with $D/T = 24.5$ from the section table.
The monosymmetry index ψ for a T-section with lipped flanges, where

D_L = depth of the lip = 76.2 mm,
D = overall depth = 465.3 mm.

$$\psi = 0.8[(2 \times 0.834) - 1][1 + (76.2/2 \times 465.3)] = 0.578.$$

The slenderness factor

$$v = \frac{1}{\left[(4(0.834)(1 - 0.834) + 0.05(88.7/26.5)^2 + 0.578^2)^{0.5} + 0.578\right]^{0.5}}$$
$$= 0.75$$

Table 14 gives

$$v = 0.769 \text{ for } \eta = 0.834 \quad \text{and} \quad \lambda/x = 88.7/26.5 = 3.34.$$

The equivalent slenderness

$$\lambda_{LT} = 1.0 \times 0.769 \times 88.7 \times 1.0 = 68.2.$$

From Table 17 for welded sections for $p_y = 265 \text{ N/mm}^2$.
Top flange thickness = 23.6 mm total

$$p_b = 154.3 \text{ N/mm}^2.$$

The buckling resistance

$$M_b = 2086.8 \times 154.3 \times 10^{-3} = 321.9 \text{ kN m}.$$

(3) Moment capacity for the top section for the y–y axis

$$M_{cy} = 265 \times 331 \times 10^{-3} = 87.7 \,\text{kN m},$$
$$Z_y = 331 \,\text{cm}^3 \text{ (from the section table).}$$

(4) Check beam in bending

(1) Vertical moment, no horizontal moment:

$$M_x = 267.5 \,\text{kN m} \quad < M_b = 321.9 \,\text{kN m}.$$

(2) Vertical moment no impact + horizontal moment:

$$\frac{M_x}{M_b} + \frac{M_y}{p_y Z_y} = \frac{215.5}{321.9} + \frac{7.89}{87.7} = 0.76 < 1.$$

(3) Vertical moment with impact + horizontal moment:

$$\frac{M_x}{M_b} + \frac{M_y}{p_y Z_y} = \frac{234.9}{321.9} + \frac{6.9}{87.7} = 0.81 < 1.$$

The crane girder is satisfactory in bending.

(5) Shear Capacity (see Section 4.62)

$$P_v = 0.6 \times 457.2 \times 9.1 \times 265 \times 10^{-3} = 611.5 \,\text{kN},$$
Maximum factored shear $= 250.4 \,\text{kN}.$

(6) Weld between channel and universal beam

The dimensions for determining the horizontal shear are shown in Figure 4.36(a). The location of the centroidal axis is taken from the *Structural Steelwork Handbook*.

Horizontal shear force in each weld

$$= \frac{250.4 \times 3603 \times 158.1}{45983 \times 10^4} = 0.31 \,\text{kN/mm}.$$

Provide 6 mm continuous fillet (weld strength $= 0.92 \,\text{kN/mm}$).

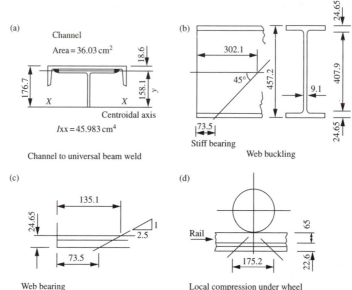

Figure 4.36 Diagram for crane beam design

(7) Web bearing and buckling

Assume a stiffened bracket 200 mm wide. The stiff bearing length allowing 3 mm clearance between the beams is 73.5 mm (see Section 4.8 and Figure 4. 36(c)).

Factored reaction = 250.4 kN
Bearing capacity = $135.1 \times 9.1 \times 265 \times 10^{-3}$
$\qquad\qquad\qquad = 325.8$ kN

∴ Satisfactory
Buckling resistance:

$$P_x = \left[\frac{200 + 0.7(407.9)}{1.4(407.9)}\right] \frac{25(1.0)(9.1)}{\sqrt{(135.1)(407.9)}}(325.8)$$
$$= 268.4\,\text{kN}$$

∴ Satisfactory

(8) Local compression under wheels

A 25 kg/m crane rail is used, and the height H_R is 65 mm. The length in bearing is shown in Figure 4.36(d):

Bearing capacity = $265 \times 9.1 \times 175.2 \times 10^{-3} = 422.4$ kN
Factored crane wheel load = $99 \times 1.6 = 158.4$ kN
∴ Satisfactory

(9) Deflection

The vertical deflection due to the static wheel load must not exceed:

$$\text{Span}/600 = 5500/600 = 9.17\,\text{mm}.$$

See Figures 4.32 and 4.33. The horizontal deflection due to crane surge must not exceed:

$$\text{Span}/500 = 5500/500 = 11\,\text{mm}.$$

The vertical deflection at the centre with the loads in position for maximum moment is

$$\delta = \frac{79.2 \times 10^3 \times 5500^3}{48 \times 205000 \times 45983 \times 10^4}\left(\frac{3(875 + 2125)}{5500} - \frac{4(875^3 + 2125^3)}{5500^3}\right)$$
$$+ \frac{5 \times 8000 \times 5500^3}{384 \times 205000 \times 45983 \times 10^4} = 4.03 + 0.18 = 4.21\,\text{mm}$$

If the loads are placed equidistant about the centre line of beam, $a = c$ is 1500 mm:

$$\delta = 4.29 + 0.18 = 4.47\,\text{mm}.$$

This gives the maximum deflection.
The horizontal deflection due to the surge loads

$$= \frac{4.29 \times 3 \times 45983}{79.2 \times 4202.5} = 1.77\,\text{mm}.$$

The crane girder is satisfactory with respect to deflection.

4.12 Purlins

4.12.1 Types and uses

The purlin is a beam and it supports roof decking on flat roofs or cladding on sloping roofs on industrial buildings.

Members used for purlins are shown in Figure 4.37. These are cold-rolled sections, angles, channels joists and structural hollow sections. Cold-rolled sections are now used on most industrial buildings.

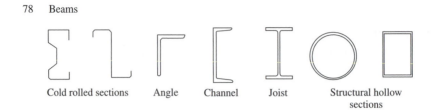

Cold rolled sections   Angle   Channel   Joist   Structural hollow sections

Figure 4.37 Section used for purlins and sheeting rails

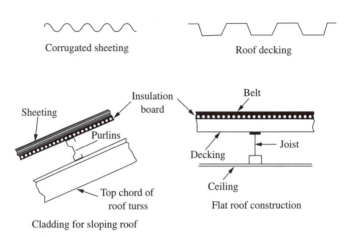

Corrugated sheeting   Roof decking

Sheeting   Insulation board   Belt   Joist

Purlins   Decking

Top chord of roof turss   Ceiling

Cladding for sloping roof   Flat roof construction

Figure 4.38 Roof material and constructions

4.12.2 Loading

Roof loads are due to the weight of the roof material and the imposed load. The sheeting may be steel or aluminium corrugated or profile sheets or decking. On sloping roofs, sheeting is placed over insulation board or glass wool. On flat roofs, insulation board, felt and bitumen are laid over the steel decking. Typical roof cladding and roof construction for flat and sloping roofs are shown in Figure 4.38.

The weight of roofing varies from 0.3 to 1.0 kN/m^2, including the weight of purlins or joists, and the manufacturer's literature should be consulted. Purlins carrying sheeting are usually spaced at from 1.4–2.0 m centres. Joists carrying roof decking can be spaced at larger centres up to 6 m or more, depending on thickness of decking sheet and depth of profile.

Imposed loading for roofs is specified in BS 6399: Part 3 in Section 4.

(1) *Flat roofs*: On flat roofs and sloping roofs up to and including 10°, where access in addition to that necessary for cleaning and repair is provided to the roof, allowance shall be made for an imposed load, including snow of 1.5 kN/m^2 measured on plan or a load of 1.8 kN concentrated. On flat roofs and sloping roofs up to and including 10°, where no access is provided to the roof other than that necessary for cleaning and repair, allowance shall be made for an imposed load, including snow of 0.6 kN/m^2 measured on plan or a load of 0.9 kN concentrated.

(2) *Sloping roofs:* On roofs with a slope greater than 10° and with no access provided to the roof other than that necessary for cleaning and repair the following imposed loads, including snow, shall be allowed for:

(a) For a roof slope of 30° or less, 0.6 kN/m^2 measured on plan or a vertical load of 0.9 kN concentrated;

(b) For a roof slope of 60° or more, no allowance is necessary. For roof slopes between 30° and 60°, a uniformly distributed load of $0.6[(60 - \alpha)/30]$ kN/m^2 measured on plan where α is the roof slope.

Wind loads are generally upward, or cause suction on all but steeply sloping roofs. In some instances, the design may be controlled by the dead-wind load cases. Wind loads are estimated in accordance with BS 6399 Part 2. The calculation of wind loads on a roof is given in Chapter 8 of this book.

4.12.3 Purlins for a flat roof

These members are designed as beams with the decking providing full lateral restraint to the top flange. If the ceiling is directly connected to the bottom flange the deflection due to imposed load may need to be limited to span/360, in accordance with Table 8 of BS 5950-1: 2000. In other cases the code states in Section 4.12.2 that the deflection should be limited to suit the characteristics of the cladding system.

4.12.4 Purlins for a sloping roof

Consider a purlin on a sloping roof as shown in Figure 4.39(a). The load on an interior purlin is from a width of roof equal to the purlin spacing S. The load is made up of dead and imposed load acting vertically downwards.

A conservative method of design is to neglect the in-plane strength of the roof, resolve the load normal and tangential to the roof surface and design the purlin for moments about the $x–x$ and $y–y$ axes (see Figure 4.39(c)). If a section such as a channel is used where the strength about the $y–y$ axis is much less than that about the $x–x$ axis, a system of sag rods to support the purlin about the weak axis may be introduced, as shown in Figure 4.39(b). The purlin is then designed as a simply supported beam for bending about the $x–x$ axis and a continuous beam for bending about the $y–y$ axis.

A more realistic and economic design results if the in-plane strength of the cladding is taken into account. The purlin is designed for bending about the $x–x$ axis with the whole vertical load assumed to cause moment.

An angle purlin bent at the full plastic moment about the $x–x$ axis is shown in Figure 4.39(d). Note that the internal resultant forces act at the centroids of the tension and compression areas. These forces cause a secondary moment about the $y–y$ axis. It is assumed in design that the sheeting absorbs this moment.

BS 5950-1: 2000 gives the classification for angles in Table 11, where limiting width/thickness ratios are given for legs. The sheeting restrains the angle member so that bending take place about the $x–x$ axis. The unsupported downward leg is in tension in simply supported purlins, but it would be in compression under uplift from wind load or near the supports in continuous purlins.

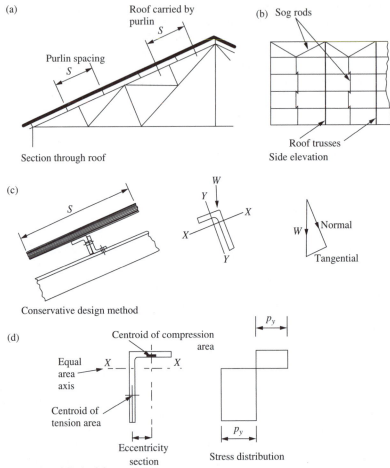

Figure 4.39 Design of purlins for a sloping roof

The moment capacity for semi-compact outstand elements and a conservative value for plastic and compact sections is:

$$M_c = p_y Z_x,$$
$$Z_x = \text{elastic modulus for the } x\text{--}x \text{ axis.}$$

4.12.5 Design of purlins to BS 5950-1: 2000, Section 4.12

The code states that the cladding may be assumed to provide restraint to an angle section or to the face against which it is connected in the case of other sections. Deflections as mentioned above are to be limited to suit the characteristics of the cladding used.

The empirical design method is set out in Section 4.12.4 of the code, and the general requirements are:

(1) The member should be of steel to a minimum of grade S275.
(2) Unfactored loads are used in the design;
(3) The span is not to exceed 6.5 m centre to centre of main supports;
(4) If the purlin spans one bay it must be connected by two fasteners at each end;
(5) If the purlins are continuous over two or more bays with staggered joints in adjacent lines, at least one end of any single-bay member should be connected by not less than two fasteners.

The rules for empirical design of angle purlins are:

(1) The roof slope should not exceed 30°.
(2) The load should be substantially uniformly distributed. Not more than 10 per cent of the total load should be due to other types of load;
(3) The elastic modulus about the axis parallel to the plane of cladding should not be less than the larger value of $W_p/1800\,\mathrm{cm}^3$ or $W_q L/2250\,\mathrm{cm}^3$, where W_p is the total unfactored load on one span (kN) due to dead and imposed load and W_q is the total unfactored load on one span (kN) due to dead minus wind load and L is the span (mm).
(4) Dimension D perpendicular to the plane of the cladding is not to be less than $L/45$. Dimension B parallel to the plane of the cladding is not to be less than $L/60$.

The code notes that where sag rods are provided the sag rod spacing may be used to determine B only.

4.12.6 Cold-rolled purlins

Cold-rolled purlins are almost exclusively adopted for industrial buildings. The design is to conform to BS 5950: Part 5, Code of Practice for Design of Cold Formed Sections. Detailed design of these sections is outside the scope of this book.

The purlin section for a given roof may be selected from manufacturer's data. Ward Building Components Ltd has kindly given permission for some of their design data to be reproduced in this book. This firm produces complete systems for purlins and cladding rails based on their cold-formed multibeam section. Full information including fixing methods and accessories is given in their Technical Handbook (12). In addition, they have produced the multibeam design software system for optimum design for their purlins and side rails.

The multibeam cold-formed section and ultimate loads for double-span purlins for a limited range of purlins are shown in Table 4.2. Notes for use of the table are:

(1) The loads tables show the ultimate loads that can be applied. The section self-weight has not been deducted. Loadings have also been tabulated that will produce the noted deflection ratio.
(2) The loads given are based on lateral restraint being provided to the top flange by the cladding,
(3) The values given are also the ultimate uplift load due to wind uplift.

Table 4.2 Ward building components cold-formed purlin—design data

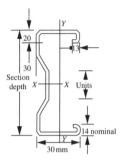

Example: Section P175150
Depth = 175 mm;thickness = 1.5 mm

Double Span Loads

Span m	Section	Depth D	Self wt Kg/m	Ult gravity kN	Ult uplift kN	Def limit L/180 kN
4.5	P145130	145	3.03	12.90	10.32	12.96
	P145145	145	3.38	15.58	12.47	14.43
	P145155	145	3.62	17.38	13.90	15.46
	P145170	145	3.97	19.93	15.94	16.89
	P175140	175	3.59	18.36	14.69	21.71
	P175150	175	3.85	20.62	16.50	23.26
	P175160	175	4.05	22.52	18.02	24.47
	P175170	175	4.36	25.01	20.01	26.32
5.0	P145130	145	3.03	11.76	9.41	10.50
	P145145	145	3.38	14.18	11.34	11.69
	P145155	145	3.62	15.80	12.64	12.52
	P145170	145	3.97	18.10	14.48	13.68
	P175140	175	3.59	16.77	13.42	17.59
	P175150	175	3.85	18.81	15.05	18.84
	P175160	175	4.05	20.51	16.41	19.84
	P175170	175	4.36	22.77	18.22	21.32

4.12.7 Purlin design examples

Example 1. Design of a purlin for a flat roof

The roof consists of steel decking with insulation board, felt and rolled-steel joist purlins with a ceiling on the underside. The total dead load is $0.9\,\text{kN/m}^2$ and the imposed load is $1.5\,\text{kN/m}^2$. The purlins span 4 m and are at 2.5 m centres. The roof arrangement and loading are shown in Figure 4.40. Use Grade S275 steel.

Dead load $= 0.9 \times 4 \times 2.5 = 9\,\text{kN}$,
Imposed load $= 1.5 \times 4 \times 2.5 = 15\,\text{kN}$,
Design load $= (1.4 \times 9) + (1.6 \times 15) = 36.6\,\text{kN}$,
Moment $= 36.6 \times 4/8 = 18.3\,\text{kN}$,
Design strength, $p_y = 275\,\text{N/mm}^2$,

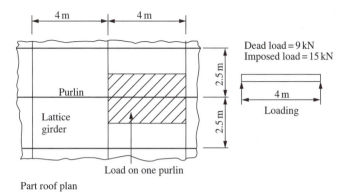

Figure 4.40 Purlin for a flat roof

Modulus required, $Z_{req'd} = 18.3 \times 10^3/275 = 66.54\,cm^3$.

Try 127×76 joist 13.36 kg/m, $Z_x = 74.94\,cm^3$, $I_x = 475.9\,cm^4$.

Deflection due to imposed load:

$$\delta = \frac{5 \times 15 \times 10^3 \times 4000^3}{384 \times 205 \times 10^3 \times 475.9 \times 10^4} = 12.81\,mm,$$

$\delta/span = 12.81/4000 = 1/312 > 1/360$,

Increase section to 127×76 joist 16.37 kg/m, $I_x = 569.4\,cm^4$.

$\delta/span = 1/373$ (satisfactory),

Purlin 127×76 joist 16.37 kg/m.

Example 2. Design of an angle purlin for a sloping roof

Design an angle purlin for a roof with slope 1 in 2.5. The purlins are simply supported and span 5.0 m between roof trusses at a spacing of 1.6 m. The total dead load, including purlin weight, is 0.32 kN/m² on the slope and the imposed load is 0.6 kN/m² on plan. Use Grade S275 steel. The arrangement of purlins on the roof slope and loading are shown in Figure 4. 41.

Dead load on slope $= 0.32 \times 5 \times 1.6 = 2.56\,kN$,
Imposed load on plan $= 0.6 \times 5 \times 1.6 \times 2.5/2.69 = 4.46\,kN$,
Design load $= (1.4 \times 2.56) + (1.6 \times 4.46) = 10.72\,kN$,
Moment $= 10.72 \times 5/8 = 6.7\,kN\,m$.

Assume that the angle bending about the x–x axis resists the vertical load. The horizontal component is taken by the sheeting.

Design strength, $p_y = 275\,N/mm^2$,
Applied moment = moment capacity of a single angle $6.7 \times 10^3 = 275 \times Z_x$

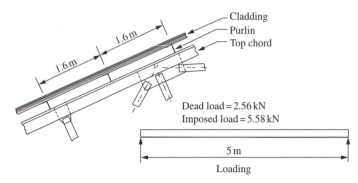

Figure 4.41 Angle purlin for a sloping roof

Elastic modulus $Z_x = 24.4\,\text{cm}^3$,
  Provide $125 \times 75 \times 8\,\text{L} \times 12.2\,\text{kg/m}$, $Z_x = 29.6\,\text{cm}^3$.

Deflection need not be checked in this case.

Example 3. Design using empirical method from BS 5950-1: 2000

Redesign the angle purlin above using the empirical method from Section 4.12.4.The purlin specified meets the requirements for the design rules.

W_p = total unfactored dead + imposed load = 7.02 kN,
W_q = total unfactored wind + dead load = 3.07 kN

$$Z_p = \frac{7.02 \times 5000}{1800} = 19.5\,\text{cm}^3,$$

$$Z_q = \frac{3.07 \times 5000}{2250} = 6.82\,\text{cm}^3,$$

Elastic modulus, $Z = 19.5\,\text{cm}^3$.
Leg length perpendicular to plane of cladding, $D = 5000/45 = 111.1$ mm,
Leg length parallel to plane of cladding, $B = 5000/60 = 83.3$ mm,
Provide $120 \times 120 \times 8\,\text{L} \times 14.7\,\text{kg/m}$, $Z_x = 29.5\,\text{cm}^3$.

Example 4. Select a cold-formed purlin to meet the above requirements

Try purlin section P145130 from Table 4.2.

Dead load on slope = $0.32 \times 5 \times 1.6 = 2.56$ kN,
Imposed load on plan = $0.6 \times 5 \times 1.6 \times 2.5/2.69 = 4.46$ kN,
Wind load = $0.7 \times 5 \times 1.6 = 5.6$ kN,
Design load (gravity) = $(1.4 \times 2.56) + (1.6 \times 4.46) = 10.72$ kN,
Design load (uplift) = $(1.0 \times 2.56) - (1.4 \times 5.6) = -5.28$ kN.

The section is satisfactory and is much lighter than angle section.

4.13 Sheeting rails

4.13.1 Types of uses

Sheeting rails support cladding on walls and the sections used are the same as those for the purlins shown in Figure 4.37.

4.13.2 Loading

Sheeting rails carry a horizontal load from the wind and a vertical one from self-weight and the weight of the cladding. The cladding materials are the same as used for sloping roofs (metal sheeting on insulation board). Wind loads are estimated using BS 6399: Part 2. For design examples in this section suitable values for wind loads will be assumed.

The arrangement of sheeting rails on the side of a building is shown in Figure 4.42(a) and the loading on the rails is shown in Figure 4.42(b). The wind may act in either direction due to pressure or suction on the building walls.

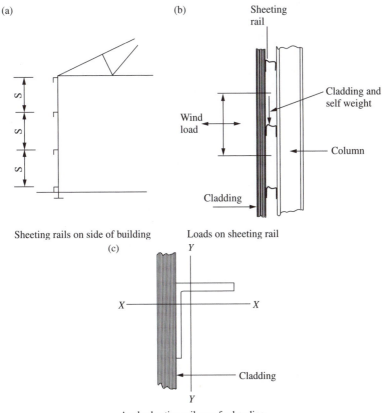

Figure 4.42 Sheeting rails: arrangement and loading

4.13.3 Design of angle sheeting rail

Sheeting rails may be designed as beams bending about two axes. It is assumed for angle sheeting rails that the sheeting restrains the member and bending takes place about the vertical and horizontal axes. Eccentricity of the vertical loading (shown in Figure 4.42(b)) is not taken into account.

The sheeting rail is fully supported on the downward leg. The outstand leg for simply supported sheeting rails is in compression from dead load and tension or compression from wind load.

The moment capacity is (see Section 4.12.4):

$$M_c = p_y Z$$

where Z is the elastic modulus for the appropriate axis.
For biaxial bending:

$$\frac{m_x M_x}{p_y Z_x} + \frac{m_y M_y}{p_y Z_y} \le 1,$$

$$\frac{m_{LT} M_{LT}}{M_b} + \frac{m_y M_y}{p_y Z_y} \le 1.$$

4.13.4 Design of angle sheeting rails to BS 5950-1: 2000

The general requirements from Section 4.12.4.1 of the code set out for purlins in Section 4.12.4 above must be satisfied. Empirical rules for design of sheeting rails are given in Section 4.12.4.3 of the BS 5950. These state that:

(1) The loading should generally be due to wind load and weight of cladding. Not more than 10 per cent should be due to other loads or due to loads not uniformly distributed.
(2) The elastic moduli for the two axes of the sheeting rail from Table 28 in the code should not be less than the following values for an angle (see Figure 4.42(c):
    (a) y–y axis—parallel to plane of the cladding:

$$Z_1 > W_1 L_1 / 2250 \, \text{cm}^3,$$

where $W_1 =$ unfactored load on one rail acting perpendicular to the plane of the cladding in kN. (This is the wind load.)
       $L_1 =$ span in millimetres, centre to centre of columns.

    (b) x–x axis—perpendicular to the plane of the cladding:

$$Z_2 > W_2 L_2 / 1200 \, \text{cm}^3,$$

where $W_2 =$ unfactored load on one railing acting parallel to the plane of the cladding in kN. (This is the weight of the cladding and rail.)
       $L_2 =$ span centre to centre of columns or spacing of sag rods where these are provided and properly supported.
(3) The dimensions of the angle should not be less than the following:
    D—perpendicular to the cladding $< L_1/45$,
    B—parallel to the cladding $< L_2/60$.
    L_1 and L_2 were defined above.

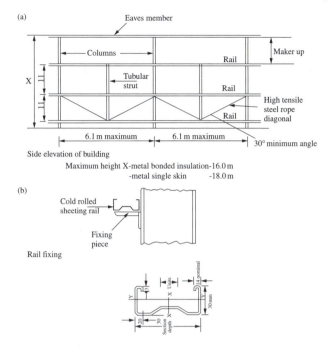

(a)

Eaves member

Columns

Rail

Maker up

X

Tubular
strut

Rail

Rail

High tensile
steel rope
diagonal

6.1 m maximum   6.1 m maximum

30° minimum angle

Side elevation of building

Maximum height X-metal bonded insulation-16.0 m
-metal single skin   -18.0 m

(b)

Cold rolled
sheeting rail

Fixing
piece

Rail fixing

Figure 4.43 Cold-formed sheeting rail system

4.13.5 Cold-formed sheeting rails

The system using cold-formed sheeting rails designed and marketed by Ward Building Components is described briefly with their kind permission.

The rail member is the Multibeam section placed with the major axis vertical. For bay widths up to 6.1 m, a single tubular steel strut is provided to support the rails at mid-span. The strut is supported by diagonal wire rope ties and the cladding system can be levelled before sheeting by adjusting the ties. The system is shown in Figure 4.43. For larger width bays, two struts are provided.

The allowable applied wind loads for a limited selection of sheeting rail spans and their ultimate loads are given in Table 4.3. Notes regarding use of the table are given below. The manufacturer's Technical Handbook should be consulted for full particulars regarding safe wind loads and fixing details for rails, support system and cladding.

Notes relating to Table 4.3 are:

(1) The loads shown are valid only when the rails and cladding are fixed exactly as indicated by the manufacturer.
(2) The loads shown are for positive external wind loads (ultimate pressure) and negative suction loads (ultimate suction).
(3) Interpolation of the ultimate loads shown is permissible on a linear basis.

4.13.6 Sheeting rail design examples

Example 1. Design of an angle sheering rail

A simply supported sheeting rail spans 5 m. The rails are at 1.5 m centres. The total weight of cladding and self weight of rail is 0.32 kN/m². The wind

Table 4.3 Ward Multibeam Cladding Rails

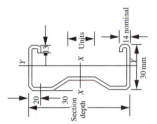

Example: Section P145170
Depth = 145 mm;thickness = 1.7 mm

Double Span Loads

Span m	Section	Depth D	Self wt Kg/m	Ult pressure kN	Ult suction kN	Def limit L/150 kN
4.5	P145130	145	3.03	12.898	10.318	*****
	P145145	145	3.38	15.583	12.456	*****
	P145155	145	3.62	17.377	13.901	*****
	P145170	145	3.97	19.928	15.942	*****
	P175140	175	3.59	18.932	14.690	*****
	P175150	175	3.85	20.624	16.499	*****
5.0	P145130	145	3.03	11.763	9.410	*****
	P145145	145	3.98	14.184	11.347	14.029
	P145155	145	3.62	15.800	12.640	15.023
	P145170	145	3.97	18.098	14.478	16.420
	P175140	175	3.59	16.722	19.418	*****
	P175150	175	3.85	18.811	15.049	*****
	P145160	175	4.05	20.513	16.410	*****
5.5	P145130	145	3.03	10.908	8.674	10.409
	P145145	145	3.38	13.012	10.410	11.594
	P145155	145	3.62	14.483	11.587	12.416
	P145170	145	3.97	16.573	13.259	13.570
	P175140	175	3.59	15.430	12.344	*****
	P175150	175	3.85	17.268	13.829	*****

***** Indicates the load to produce a deflection of Span/150 exceeds ultimate UDL
capacity

loading on the wall is $\pm 0.5 \, \text{kN/m}^2$. The wind load would have to be carefully
estimated for the particular building and the maximum suction and pressure
may be different. The sheeting rail arrangement is shown in Figure 4.44(a).
Use Grade S275 steel.

Vertical load $= 0.32 \times 1.5 \times 5 = 2.4 \, \text{kN}$,
Horizontal load $= 0.5 \times 1.5 \times 5 = 3.75 \, \text{kN}$.

The loading is shown in Figure 4.44(b).
The load factor $\gamma_f = 1.4$ for a wind load acting with dead load only. (Table 2
of BS 5950-1: 2000).

Factored vertical moment, $M_{cx} = 1.4 \times 2.4 \times 5/8 = 2.10 \, \text{kN m}$,
Factored horizontal moment, $M_{cy} = 1.4 \times 3.75 \times 5/8 = 3.28 \, \text{kN m}$.

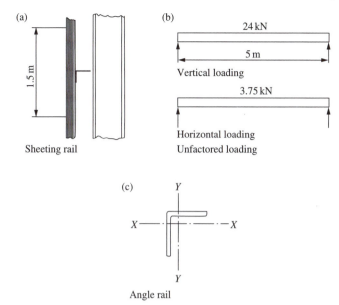

Figure 4.44 Angle sheeting rail

Design strength, $p_y = 275\,\text{N/mm}^2$.

Try $100 \times 100 \times 10\,\text{L}$ where $Z = 24.6\,\text{cm}^3$.

The moment capacity:

$$M_b = M_{cy} = 0.8 \times 275 \times 24.6 \times 10^{-3} = 5.41\,\text{kN m}.$$

The biaxial bending interaction relationship:

$$\frac{M_x}{M_b} + \frac{M_y}{M_{cy}} = \frac{2.1}{5.41} + \frac{3.28}{5.41} = 0.99 < 1.0.$$

Provide $100 \times 100 \times 10\,\text{L} \times 15\,\text{kg/m}$.
For the outstand leg, $b/t = 10$ compact (Table 11).

Example 2. Design using empirical method from BS 5950-1: 2000

Redesign the angle sheeting rail above using the empirical method from BS 5950.

Unfactored wind load $W_1 = 3.75\,\text{kN}$.
Elastic modulus

$$Z_1 = Z_y = 3.75 \times 5000/2250 = 8.33\,\text{cm}^3.$$

Unfactored dead load $W_2 = 2.4\,\text{kN}$.

Elastic modulus

$$Z_2 = Z_x = 2.4 \times 5000/1200 = 10.0\,\text{cm}^3.$$

Dimensions specified are to be

D—perpendicular to cladding $< 5000/45 = 111.1\,\text{mm}$,
B—parallel to cladding $< 5000/60 = 83.3\,\text{mm}$.

$120 \times 120 \times 8\,\text{L}$ is the smallest angle to meet all the requirements.

Example 3. Select a cold-rolled sheeting rail to meet the following requirements

Wind load $= \pm 0.5\,\text{kN/m}^2$,
Span $= 5.0\,\text{m}$,
Spacing $= 1.5\,\text{m}$.

Try cladding rail section P145130 from Table 4.3.

Horizontal load $= 0.5 \times 1.5 \times 5 = 3.75\,\text{kN}$,
Design load (pressure or suction) $= 1.4 \times 3.75 = 5.25\,\text{kN}$.

This section is satisfactory. (See Figure 4.43 for the rail support system.)

Problems

4.1 A simply supported steel beam of 6.0 m span is required to carry a uniform dead load of 40 kN/m and an imposed load of 20 kN/m. The floor slab system provides full lateral restraint to the beam. If a 457×191 UB 67 of Grade S275 steel is available for this purpose, check its adequacy in terms of bending, shear and deflection.

4.2 The beam carries the same loads as in Problem 4.1, but no lateral restraint is provided along the span of the beam. Determine the new size of universal beam required.

4.3 A steel beam of 8.0 m span carries the loading as shown in Figure 4.45. Lateral restraint is provided at the supports and the point of concentrated load (by cross beams). Using Grade S275 steel, select a suitable universal beam section to satisfy bending, shear and the code's serviceability requirements.

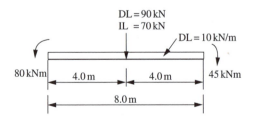

Figure 4.45

4.4 It is required to design a beam with an overhanging end. The dimension and loading are shown in Figure 4.46. The beam has torsional restraints at the supports but no intermediate lateral support. Select a suitable universal beam using Grade S275 steel.

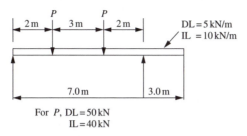

Figure 4.46

4.5 A 610 × 229 UB 125 is used as a roof beam. The arrangement is shown in Figure 4.47 and the beam is of Grade S275 steel and fully restrained by the roof decking. Check the adequacy of the section in bending and shear and the web in buckling and crushing.

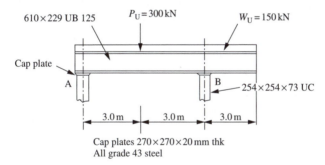

Figure 4.47

4.6 The part floor plan for the internal panel of an office building is shown in Figure 4.48. The floor is precast concrete slabs 125 mm thick supported on

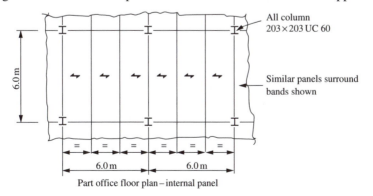

Figure 4.48

steel beams. The following loading data may be used:

125 mm concrete slab $= 3.0\,\text{kN/m}^2$,
Screed finishes $= 1.0\,\text{kN/m}$,
Partition $= 1.0\,\text{kN/m}^2$,
Imposed load $= 3.0\,\text{kN/m}^2$.

Design the floor beams, assuming that the self weight of main beams and secondary beams may be taken as 0.5 and 1.0 kN/m run, respectively.

4.7 A simply supported girder is required to span 7.0 m. The total load including self-weight of girder is 130 kN/m uniformly distributed. The overall depth of the girder must not exceed 500 mm and a compound girder is proposed. If the compression flange has adequate lateral restraint and the two flange plates are not curtailed, carry out the following work:

(a) Check that a section consisting of 457×191 UB 98 and two No. 15×250 flange plates is satisfactory;
(b) Determine the weld size required for the plate-to-flange weld at the point of maximum shear;
(c) If the girder is supported on brackets at each end with a stiff bearing length of 80 mm, check the web shear, buckling and crushing.

4.8 A simply supported crane girder for a 200 kN (working load) capacity electric overhead crane spans 7 m. The maximum static wheel loads from the end carriage are shown in Figure 4.49. It is proposed to use a crane girder consisting of 533×210 UB 122 and $305 \times 89 \times 42$ kg/m Channel.

The weight of the crab is 40 kN and the self-weight of the girder may be taken as 15 kN. Check the adequacy of the girder section.

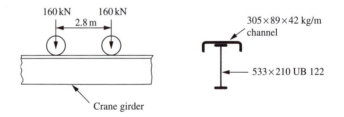

Figure 4.49

4.9 A factory building has combined roof and crane columns at 8 m centres. It is required to install an electric overhead travelling crane. Design the crane girder using simply supported spans between columns.
The crane data are as follows:

Hook load $= 150\,\text{kN}$,
Span of crane $= 15\,\text{m}$,
Weight of crane bridge $= 180\,\text{kN}$,
Weight of crab $= 40\,\text{kN}$,
No. of wheels in end carriage $= 2$,

Wheels centres in end carriage = 3 m,
Minimum hook approach = 1 m.

4.10 Select a suitable size for a simply supported cold-rolled purlin. The purlin
span is 5.0 m and the spacing is 1.8 m. The total dead load and imposed load
on plan are 0.22 and 0.6 kN/m^2, respectively. Use Table 4.2 in the design.
Redesign the purlin using the rules from BS 5950-1: 2000.

5

Plate girders

5.1 Design considerations

5.1.1 Uses and construction

Plate girders are used to carry larger loads over longer spans than are possible with rolled universal or compound beams. They are used in buildings and industrial structures for long-span floor girders, heavy crane girders and in bridges.

Plate girders are constructed by welding steel plates together to form I-sections. A closed section is termed a 'box girder'. Typical sections, including a heavy fabricated crane girder, are shown in Figure 5.1(a).

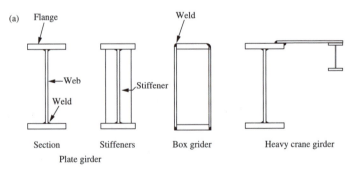

(a)

Section
Plate girder

Stiffeners

Box grider

Heavy crane girder

Sections for fabricated girders

(b)

Side elevation of a plate girder

Figure 5.1 Plate girder construction

To be competitive and cost effective, the web of a plate girder is made relatively thin compared to rolled section, and stiffeners are introduced to prevent buckling either due to compression from bending or shear. Tension field action is utilized to increase the shear buckling resistance of the thin web. Stiffeners are also required at load points and supports. Thus the side elevation of a plate girder has an array of stiffeners as shown in Figure 5.1(b).

5.1.2 Depth and breadth of flange

The depth of a plate girder may be fixed by headroom requirements but it can often be selected by the designer. The depth is usually made from one-tenth to one-twelfth of the span. The breadth of flange plate is made about one-third of the depth.

The deeper the girder is made, the smaller are the flange plates required. However, the web plate must then be made thicker or additional stiffeners provided to meet particular design requirements. A method to obtain the optimum depth is given in Section 5.3.4. A shallow girder can be very much heavier than a deeper girder in carrying the same loads.

5.1.3 Variation in girder sections

Flange cover plates can be curtailed or single flange plates can be reduced in thickness when reduction in bending moment permits. This is shown in Figure 5.2(a). In the second case mentioned, the girder depth is kept constant throughout.

For simply supported girders, where the bending moment is maximum at the centre, the depth may be varied, as shown in Figure 5.2(b). In the past, hog-back or fish-belly girders were commonly used. In modem practice with

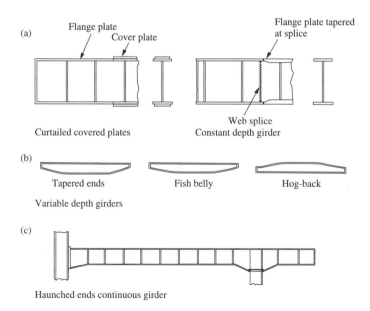

Figure 5.2 Variation in plate girder sections

automatic methods of fabrication, it is more economical to make girders of uniform depth and section throughout.

In rigid frame construction and in continuous girders, the maximum moment occurs at the supports. The girders may be haunched to resist these moments, as shown in Figure 5.2(c).

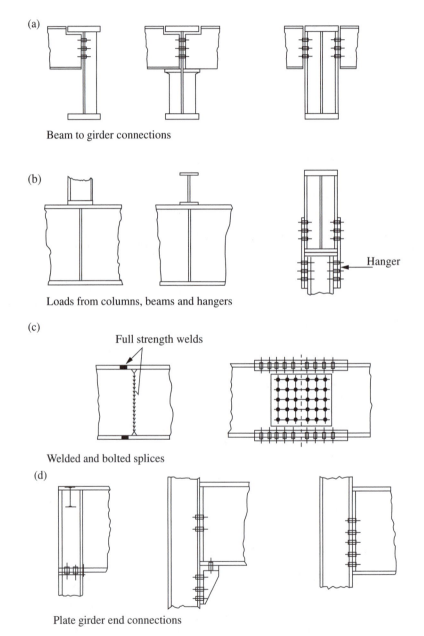

(a)

Beam to girder connections

(b)

Loads from columns, beams and hangers

(c)

Full strength welds

Welded and bolted splices

(d)

Plate girder end connections

Figure 5.3 Plate girder connections and splices

5.1.4 Plate girder loads

Loads are applied to plate girders through floor slabs, floor beams framing into the girder, columns carried on the girder or loads suspended from it through hangers. Some examples of loads applied to plate girders through secondary beams, a column and hanger are shown in Figure 5.3.

5.1.5 Plate girder connections and splices

Typical connections of beams and columns to plate girders are shown in Figures 5.3(a) and (b). Splices are necessary in long girders. Bolted and welded splices are shown in Figure 5.3(c) and end supports in Figure 5.3(d).

5.2 Behaviour of a plate girder

5.2.1 Girder stresses

The stresses from moment and shear for a plate and box girder in the elastic state are shown in Figure 5.4. The flanges have uniform direct stresses and the web shear and varying direct stress.

Plate and box girders are composed of flat plate elements supported on one or both edges and loaded in plane by bending and shear. The way in which the girder acts is determined by the behaviour of the individual plates.

5.2.2 Elastic buckling of plates

The components of the plate and box girder under stress can be represented by the four plates loaded as shown in Figure 5.5. The way in which the plates buckle and their critical buckling stresses depend on the edge conditions, dimensions and loading. The buckled plate patterns are also shown in the figure.

In all cases, the critical buckling stress can be expressed by the equation:

$$p_{cr} = K \frac{\pi^2 E}{12\left(1 - v^2\right)} \left(\frac{t}{b}\right)^2$$

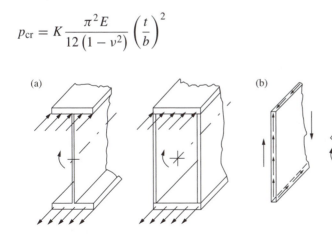

(a)   (b)

Flange bending stresses   Web shear and bending stresses

Figure 5.4 Stresses in plate and box girders

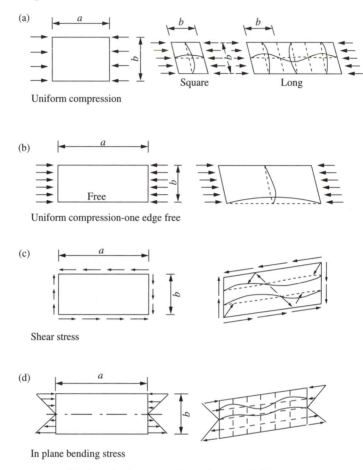

Figure 5.5 Elastic buckling of plates

where K is the buckling coefficient that depends on the ratio of plate length to width a/b (the edge conditions and loading case), E the Young's modulus, v the Poisson's ratio and t the plate thickness.

Some values of K for the four plates are shown in Figure 5.6. Note that the plate length a shown is also the stiffener spacing on a plate girder.

The critical stress depends on the width/thickness ratio b/t. Limiting values of b/t, where the critical stress equals the yield stress, are also shown in Figure 5.6. These values are for Grade S275 steel for plate up to 16 mm thick, where the yield stress $p_y = 275\,\text{N/mm}^2$. The values form the basis for Class 3 semi-compact section classification given in Table 11 in BS 5950.

The web plates of girders are subjected to combined stresses caused by direct bending stress and shear. An interaction formula is used to obtain critical stress combinations. Discussion of this topic is outside the scope of this book, where simplified design procedures given in the code are used. The reader should consult references (13) and Annex H of BS5950-1: 2000.

Plate and load	$\dfrac{\text{Length}}{\text{Width}} = \dfrac{a}{b}$	Buckling coeffcient, K	Limiting value of b/t for $p_{cr}=$ yield stress
(plate with longitudinal compression on long edges)	1.0 5.0	4.0 4.0	52.9 51.9
(plate with one edge free)	1.0 ∞	1.425 0.425	10.3 16.0
(plate with transverse load)	1.0 ∞	9.35 5.35	78.1 60.0
(plate in shear)	1.0	25.6 minimum 23	131.3 124.4

All edges simply supported except as noted

Figure 5.6 Buckling coefficients and limiting values of width/thickness ratios

(a)   (b)

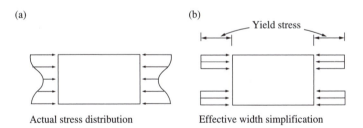

Actual stress distribution   Effective width simplification

Yield stress

Figure 5.7 Post-buckling strength: plate in compression

5.2.3 Post-buckling strength of plates

(1) Plates in compression

The plate supported on two long edges shown in Figure 5.7(a) can support more load on the outer parts following buckling of the centre portion. The behavior can be approximated by assuming that the load is carried by strips at the edge, as shown in Figure 5.7(b). The load is considered to be carried on an effective width of plate. This effective section principle is now used in the design of thin plate members which are classified as Class 4 slender sections (see Section 3.6 of the code).

A plate supported on one long edge buckles more readily than the plate above and the strength gain is not as great. Stiffeners increase the load that can be carried (see Figure 5.6).

(2) Plates in edge bending

These plates can sustain load in excess of that causing buckling. Longitudinal stiffeners in the compression region are very effective in increasing the load that can be carried. Such stiffeners are commonly provided on deep plate girders used in bridges (see Figure 5.5). However, it is not so commonly found in building steelworks and hence, longitudinal stiffeners are outside the scope of BS 5950.

(3) Plates in shear

A strength gain is possible with plates in shear where tension field action is considered. Thin unstiffened plates cannot carry much load after buckling. Referring to Figure 5.6, the critical buckling stress is increased if stiffeners are added. However, the stiffened plate can carry more loads after buckling in the diagonal tension field, as shown in Figure 5.8. The flanges, stiffeners and tension field now act like a truss (14).

If the bending strength of the flanges is ignored, the tension field develops between the stiffeners, as shown in Figure 5.8(a). If the flange contribution is included, the tension field spreads as shown in Figure 5.8(b). Failure in the girder panel occurs when plastic hinges form in the flanges and a yield zone in the web, as shown in Figure 5.8(c).

Design formulae based on theoretical and experimental work have been developed to take tension field action into account. The design method in the code also includes the flange contribution. The resistance of the web is thus the sum of the elastic buckling strength, the tension field and the flange strength.

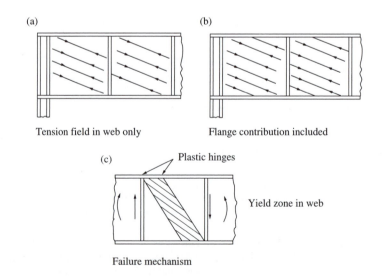

Figure 5.8 Tension field action and failure mechanism

Note that in the internal panels tension fields in adjacent panels support each other. In the end panels, the end post must be designed as a vertical beam supported by the flanges to carry the tension field (see Figures 5.8(a) and (b)). Expressions have been derived for loads on the end post.

5.3 Design to BS 5950: Part 1

5.3.1 Design strength

The strength of the thin web may be higher than that of the thicker flange due to the thickness requirements, for example, S275 steel has a design strength of 275 N/mm^2 when less than 16 mm thick and 265 N/mm^2 when greater than 16 mm thick.

The code requires that if the web strength is greater than the flange strength ($p_{yw} \geq p_{yf}$), the flange strength should be used in all calculations, including section classification, except those for shear or forces transverse to the web where the web strength may be used. If the web strength is less than the flange strength ($p_{yw} \leq p_{yf}$), both strengths may be used when considering moment or axial force, but the web strength should be used in all calculations involving shear or forces transverse to the web.

5.3.2 Classification of girder cross-sections

The classification of cross-sections from Section 3.5 of BS 5950: Part 1 was given in Section 4.3 (beams). The limiting proportions for flanges and webs for built-up sections from Table 11 in the code are given in Figure 5.9. The limits for welded sections are lower than those for rolled sections because welded sections have more severe residual stresses and fabrication errors can also adversely affect behaviour. (The reader should refer to the code for treatment of Class 4 slender cross-sections.)

5.3.3 Moment capacity

If the depth/thickness ratio d/t for the web is less than or equal to 62ε, the web is not susceptible to shear buckling and the moment capacity is determined in the same way as for restrained beams given in Section 4.4.2. The stress distribution is shown in Figure 5.10(a).

If the depth/thickness ratio d/t for the web is greater than 62ε, the web is susceptible to shear buckling. The post-buckled shear resistance of the web is defined as the simple shear buckling resistance, $V_w = dt q_w$. The shear buckling strength q_w is given in Table 21 or Annex H.1 of the code and depends on the d/t of the web and a/d of the web panel. When the applied shear reaches this level, the web will already buckle. Although the web is still capable of carrying further shear in its buckled state, its ability to take part in resisting bending moment or longitudinal compression will be reduced. Therefore, the moment capacity of the section will depend on the level of applied shear and may be obtained using one of the following methods:

(1) *Low shear*: If the applied shear is less than or equal to 60 per cent of the simple shear buckling resistance V_w, then it will not cause shear buckling

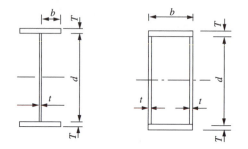

Type of element		Class of section		
		Class 1 Plastic	Class 2 Compact	Class 3 Semi-compact
Outstand element of compression flange	$\dfrac{b}{T}$	8 ε	9 ε	13 ε
Internal element of compression flange	$\dfrac{b}{T}$	28 ε	32 ε	40 ε
Web with neutral axis at mid depth	$\dfrac{b}{T}$	80 ε	100 ε	120 ε

$$\varepsilon = (275/p_y)^{0.5}$$

Figure 5.9 Classification of girder cross-sections

and the moment capacity is determined in the usual way as for restrained beams,

(2) *High shear—flange only method*: If the applied shear is greater than 60 per cent V_w, the web is designed for shear only and the flanges are not Class 4 slender, then the moment capacity may be obtained by assuming that the moment is resisted by the flanges alone with each flange subject to a uniform stress not exceeding p_{yf}.

(3) *High shear—general method*: If the applied shear is greater than 60 per cent V_w and the moment does not exceed the low shear moment capacity given in (a), then the moment capacity may be based on the capacity of the flanges plus the capacity of the web. Checks on the web contribution should be carried out to Annex H of the code.

Only method (2), i.e. flange-only method will be considered further in this book. The stress distribution in bending for this case is shown in Figure 5.10(b). The moment capacity for a girder with laterally restrained compression flange is:

$$M_c = BT(d + T)p_{yf}$$

where B is the flange, T the flange thickness and d the web depth.

For cases where the compression flange is not restrained, lateral torsional buckling may occur. This is treated in the same way that was set out for beams in Section 4.5. The bending strength P_b for welded sections is taken from Table 17 in the code.

(a)

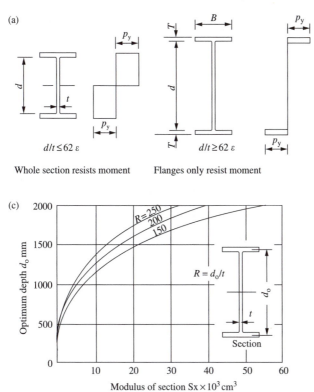

$d/t \leq 62\,\varepsilon$

Whole section resists moment

$d/t \geq 62\,\varepsilon$

Flanges only resist moment

(c)

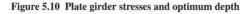

Optimal depth design chart

Figure 5.10 Plate girder stresses and optimum depth

5.3.4 Optimum depth

The optimum depth based on minimum area of cross-section may be derived as follows. This treatment applies to a girder with restrained compression flange for a given web depth/thickness ratio. Define terms:

d_0 = distance between centres of flanges,
  = d, clear depth of web approximately,
R = ratio of web depth/thickness = d_0/t,
A_f = area of flange,
S = plastic modulus based on the flanges only,
  = $A_f d_0$,
$A_f = S/d_0$,
A_w = area of web = $d_0 t = d_0^2/R$,
A = total area = $2S/d_0 + d_0^2/R$.

Differentiate with respect to d_0 and equate to zero to give

$$d_0 = (RS)^{1/3}$$

Curves drawn for depth d_0 against plastic modulus S for values of R of 150, 200 and 250 are shown in Figure 5.10(b). For the required value of

$$S = M/p_{yf}$$

the optimum depth d_0 can be read from the chart for a given value of R, where M is the applied moment.

5.3.5 Shear buckling resistance and web design

(1) Minimum thickness of web

This is given in Section 4.4.3 of BS 5950: Part 1. The following two conditions must be satisfied for webs with intermediate transverse stiffeners:

(i) Serviceability to prevent damage in handling:

Stiffener spacing $a > d$: $t \geq d/250$

Stiffener spacing $a \leq d$: $t \geq (d/250)(a/d)^{0.5}$

(ii) To avoid the flange buckling into the web. This type of failure has been observed in girders with thin webs:[13]

Stiffener spacing $a > 1.5d$: $t \geq (d/250)(p_{yf}/345)$

Stiffener spacing $a \leq 1.5d$: $t \geq (d/250)(p_{yf}/455)^{0.5}$

where d is the depth of web, t the thickness of web and p_{yf} the design strength of compression flange.

(2) Design for shear buckling resistance

The shear buckling resistance of the thin webs with $d/t > 62\varepsilon$ is covered in Section 4.4.5 of BS 5950: Part 1 and applies to webs carrying shear only. Those that are used to carry bending moment and/or axial load in addition to shear should be designed to Annex H of the code. Thin webs with intermediate stiffeners may be designed either using the simplified or more exact method.

In the simplified method, it is assumed that the flanges play no part in resisting the shear. The shear buckling resistance V_b of the thin web with intermediate transverse stiffeners should be based on the simple shear buckling resistance V_w as given in Section 4.4.5.2 of the code as:

$$V_b = V_w = dtq_w$$

where q_w is the post-buckled shear buckling strength assuming tension field action and is given in Tables 21 or Annex H.1 of the code. Unlike the old code, the critical shear buckling resistance before utilizing tension field action V_{cr} is now expressed as equation in Cl.4.4.5.4 or Annex H.2 of the code (see equations given in Section 5.3.7(4)).

The more exact method assumes that the flanges can play a part in resisting the shear. The stress in the flange due to axial load and/or bending moment as well as the strength of the flange must therefore be considered.

If the flange is fully stressed ($f_f = p_{yf}$) then the shear buckling resistance is the same as for the simplified method.

If the flanges are not fully stressed ($f_f \leq p_{yf}$), the shear buckling resistance may be increased to:

$$V_b = V_w + V_f \quad \text{but} \quad V_b \leq P_v$$

and V_f, the flange-dependent shear buckling resistance is given by:

$$V_f = \frac{P_v \, (d/a) \left[1 - \left(f_f / p_{yf} \right)^2 \right]}{1 + 0.15 \left(M_{pw} / M_{pf} \right)}$$

where, f_f is the mean longitudinal stress in the flange due to moment, $M/(d_0 BT)$, M_{pf} the plastic moment capacity of the flange, $p_{yf} BT^2/4$, M_{pw} the plastic moment capacity of the web, $p_{yw} t d^2/4$, p_{yf} the design strength of the flange and M the applied moment.

The dimensions B, T, t, d and d_0 defined above are shown in Figure 5.10.

5.3.6 Stiffener design

Two main types of stiffeners used in plate girders are:

(1) *Intermediate transverse web stiffeners*: These divide the web into panels and prevent the web from buckling due to shear. They also have to resist direct forces from tension field action and possibly external loads acting as well.
(2) *Load carrying and bearing stiffeners*: These are required at all points where substantial external loads are applied through the flange and at supports to prevent local buckling and crushing of the web.

The stiffeners at the supports are also termed 'end posts'. The design of the end posts to provide end anchorage for tension field action to develop in the end panel is dealt with in Section 5.3.9. Note that other special-purpose web stiffeners are defined in BS 5950: Part 1 in Section 4.5.1.1. Only the types mentioned above will be discussed in this book.

5.3.7 Intermediate transverse web stiffeners

Transverse stiffeners may be placed on either one or both sides of the web, as shown in Figure 5.11. Flats are the most common stiffener section used. The requirements and design procedure are set out in Section 4.4.6 of BS 5950: Part 1. Only stiffeners not subjected to any external loads or moments are considered here. The code should be consulted for design of stiffeners subjected to external loads or moments. The design process is:

(1) Spacing

This depends on:

 (i) minimum web thickness *(see Section 5.3.5(1))*;
(ii) web shear buckling resistance required. The closer the spacing is, the greater the shear buckling resistance *(see Section 5.3.5(2))*.

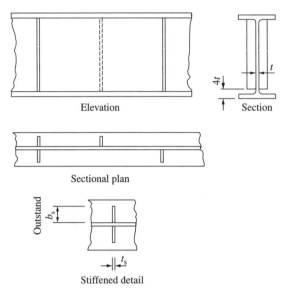

Elevation   Section

Sectional plan

Outstand

Stiffened detail

Figure 5.11 intermediate transverse web stiffeners

(2) Outstand

This is given in Section 4.5.1.2 of the code. The outstand should not exceed $19t_s\varepsilon$ (see Figure 5.11), where t_s is the thickness of stiffener and ε equals $(275/p_y)^{0.5}$.

When the outstand is between $13t_s\varepsilon$ and $19t_s\varepsilon$, the design is to be based on a core effective section with an outstand of $13t_s\varepsilon$.

(3) Minimum stiffness

Transverse stiffeners not subjected to any external loads or moments should have a second moment of inertia I_s about the centerline of the web not less than I_s given by:

$$\text{for } a/d \geq \sqrt{2}: \quad I_s = 0.75dt_{min}^3$$
$$\text{for } a/d < \sqrt{2}: \quad I_s = 1.5(d/a)^2dt_{min}^3$$

where a is the actual stiffener spacing, d the depth of web and t_{min} the minimum required web thickness for actual stiffener spacing a.

Note that additional stiffness for external loading are stipulated in Section 4.4.6.5 of the code is required where stiffeners are subject to lateral loads or to moments due to eccentricity of transverse loads relative to the web. No increase is needed where transverse loads are in line with the web.

(4) Buckling resistance

This check is only required for intermediate stiffeners in webs when tension field action is utilized. The stiffener should be checked for buckling for a force:

$$F_q = V - V_{cr} < P_q$$

where V is the maximum shear in the web panel adjacent to the stiffener, V_{cr} the critical shear buckling resistance of the same web panel given by the following:

if $V_w = P_v$	$V_{cr} = P_v$
if $P_v > V_w > 0.72P_v$	$V_{cr} = (9V_w - 2P_v)/7$
if $V_w \leq 0.72P_v$	$V_{cr} = (V_w/0.9)^2/P_v$

where P_v is the shear capacity of the web panel $= 0.6P_y dt$ and P_q the buckling resistance of the intermediate web stiffener (see Section 6.3.8).

(5) Connection to web of intermediate stiffeners

The connection between each plate and the web is to be designed for a shear of not less than:

$$t^2/(5b_s)\,(\text{kN/mm})$$

where t is the web thickness (mm) and b_s the outstand of the stiffener (mm).

The code states that intermediate stiffeners that are not subject to external forces or moments may be cut-off at about $4t$ above the tension flange. The stiffeners should extend to the compression flange but need not be connected to it (see Figure 5.11).

5.3.8 Load carrying and bearing stiffeners

Load carrying and bearing stiffeners are required to prevent local buckling and crushing of the web due to concentrated loads applied through the flange when the web itself cannot support the load. The capacity of the web alone in buckling and bearing was discussed in the earlier chapter in Sections 5.8.1 and 5.8.2, respectively.

The design procedure for these stiffeners is set out in Section 4.5 of BS 5950: Part 1. The process is as follows:

(1) Outstand

This is the same as set out for intermediate stiffeners in Section 5.3.7 (2).

(2) Buckling resistance of stiffeners

This is set out in Section 4.5.3.3 of BS 5950: Part 1 (see Figure 5.12(a)). The stiffener is designed as a 'cruciform' strut of cross-sectional area A_s at the centre of the girder where A_s is the area of stiffener plus 15 times the web thickness on either side of the centre line of the stiffener ($= 2b_s t_s + 30t^2$ where b_s the density stiffener outstand, t_s the stiffener thickness and t the web thickness).

The radius of gyration is taken about the centroidal axis of the strut area parallel to the web. The effective length L_e to be used in calculating the slenderness ratio of the stiffener acting as the strut is:

(a) Intermediate transverse stiffeners:

$$L_e = 0.7L.$$

(b) Load carrying stiffeners where the flange through which the load is applied is restrained against lateral movement is:
    (i) where the flange is restrained against rotation in the plane of the stiffener by other elements:

$$L_e = 0.7L.$$

    (ii) where the flange is not so restrained:

$$L_e = 1.0L.$$

    where L is the length of stiffener.

Note that the code states that if no effective lateral restraint is provided the stiffener should be designed as part of the compression member applying the load.

The design strength p_y from Table 9 of the code is the minimum for the web or stiffener. The reduction of 20 N/mm^2 referred to in the code for welded construction should not be applied unless the stiffeners themselves are welded sections (see Clause 4.5.3.3 in the code).

The compressive strength p_c is taken from Table 24(c) of the code. The buckling resistance is:

    for intermediate stiffener $P_q = p_c A_s > F_q,$
    for load carrying stiffener $P_x = p_c A_s > F_x$

where F_q is the intermediate stiffener force (see Section 6.3.7(4) above) and F_x the external load or reaction.

If the load carrying stiffener also acts as an intermediate web stiffener the code states that it should be checked for the effect of combined loads due to F_q and F_x in accordance with Clause 4.4.6.6 of the code.

(3) Bearing resistance

This is set out in Section 4.5.2.2 of BS 5950: Part 1. Bearing stiffeners should be designed for the applied force F_x minus the bearing capacity of the unstiffened web. The area of stiffener $A_{s.net}$ in contact with the flange is the net cross-sectional area after allowing for cope holes for welding. The bearing capacity P_s of the stiffener is given by:

$$P_s = A_{s.net} p_y.$$

The area A is shown in Figure 5.12(c). Note that the stiffener has been coped at the top to clear the web/flange weld.

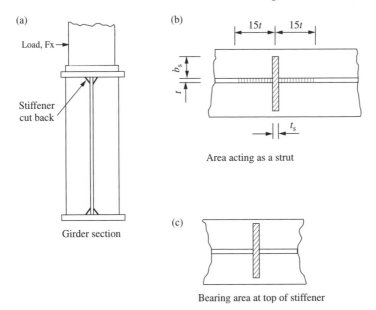

(a)

Load, Fx →

Stiffener
cut back

Girder section

(b)

b_s

t

15t 15t

t_s

Area acting as a strut

(c)

Bearing area at top of stiffener

Figure 5.12 Load-bearing stiffeners

(4) Web check between stiffeners

It may be necessary to check the compression edge of the web if loads are applied to it direct or through a flange between web stiffeners. A procedure to make this check is set out in Section 4.5.3.2 of BS 5950: Part 1. (The reader is referred to the code.)

5.3.9 End-post design

End anchorage should be provided to carry the longitudinal anchor force H_q representing the longitudinal component of the tension field at the end panel of the web with intermediate transverse stiffeners. The end post of a plate girder is provided for this purpose, and may consist of a single or twin stiffeners, as shown on Figure 5.13. The design procedure is set out in Sections 4.4.5.4 and Annex H.4 of BS 5950: Part 1. This is summarized as follows:

(1) Sufficient shear buckling resistance is available without having to utilize tension field action. Design the end post as a load carrying and bearing stiffener as set out in Section 5.3.7.
(2) End panel and internal panels are designed utilizing tension field action. In addition to carrying the reaction, the end post must be designed as a beam spanning between the flanges. The two cases shown in the figure are discussed below.

Further references should be made to the code for the case where the interior panels are designed utilizing tension field action but the end panel is not.

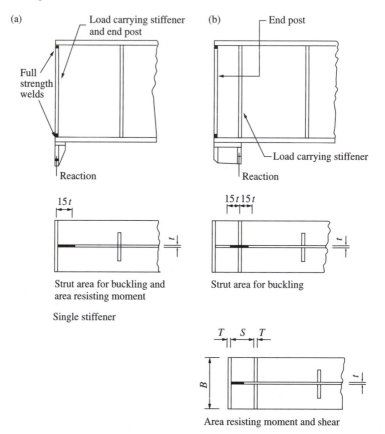

Figure 5.13 End-post design

(1) Single stiffener end-post (see Figure 5.13(a))

The single stiffener end-post also acts as both load carrying and bearing stiffener. It must be connected by full-strength welds to the flanges. The design is made for:

(i) compression due to the vertical reaction, and
(ii) in-plane bending moment M_{tf} due to the anchor force H_q.

(2) Twin stiffener end-post (see Figure 5.13(b))

The inner stiffener carries the vertical reaction from the girder. It is checked for bearing at the end and for buckling at the centre (see Section 5.3.7).

The end-post is checked as a vertical beam spanning between the flanges of the girder, with the stiffeners forming the flanges of the beam. It is designed to resist a shear force R_{tf} and a moment M_{tf} due to the longitudinal component of the tension field anchor force H_q. The moment M_{tf} induces a tension in the inner stiffener and a compression in the outer end stiffener as these form the lower and upper flanges of the vertical beam. This force F_{tf} is equal to the moment divided by the 'depth S' of the vertical beam. Thus, the stiffener must also be designed to resist this force, plus any force arising from the reaction of the plate girder.

The expressions to derive the shear, moment and compressive force given in the code are:

Shear $R_{tf} = 0.75 H_q$
Moment $M_{tf} = 0.15 H_q d$
Force $F_{tf} = M_{tf}/S$

The anchor force H_q from the tension field:

(i) if the web is fully loaded in shear $(F_v \geq V_w)$

$$H_q = 0.5 dt p_y (1 - V_{cr}/P_v)^{0.5}$$

(ii) if the web is not fully loaded in shear $(F_v < V_w)$

$$H_q = 0.5 dt p_y \left[\frac{F_v - V_{cr}}{V_w - V_{cr}} \right] (1 - V_{cr}/P_v)^{0.5}$$

where, d is the depth of the web, F_v the maximum shear force, P_v the shear capacity, t the web thickness, V_{cr} the critical shear buckling resistance and V_w the simple shear buckling resistance.

The shear capacity of the end-post is:

$$P_v = 0.6 p_y S t$$

where S is the length of web between stiffeners and t the web thickness. The shear capacity P_v must exceed the shear from the tension field R_{tf}.

The moment capacity of the end-post at the centre of the girder, assuming that the flanges resist the whole moment, is:

$$M_{cx} = p_y B T (S + T)$$

where B is the stiffener width and T the stiffener thickness. Note that proportions should be selected so that the plates selected are Class 3 semi-compact as a minimum requirement. The moment capacity M_{cx} must exceed the moment due to tension field action M_{tf}.

The welds between the stiffener and web must be designed to carry the reaction and the shear from the end-post beam action.

The application of the design procedure is given in the example in Section 6.4.

5.3.10 Flange to web welds

Fillet welds are used for the flange to web welds (see Figure 5.14). The welds are designed for the horizontal shear per weld:

$$= F A y / 2 I_x$$

where F is the applied shear, A the area of flange, y the distance of the centroid of A from the centroid of the girder and I_x the moment of inertia of the girder about the x–x axis.

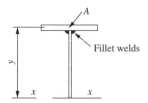

Figure 5.14 Flange-to-web weld

The fillet weld can be intermittent or continuous, but continuous welds made by automatic welding are generally used.

5.4 Design of a plate girder

A simply supported plate girder has a span of 12 m and carries two concentrated loads on the top flange at the third points consisting of 450 kN dead load and 300 kN imposed load. In addition, it carries a uniformly distributed dead load of 20 kN/m, which includes an allowance for self-weight and an imposed load of 10 kN/m. The compression flange is fully restrained laterally. The girder is supported on a heavy stiffened bracket at each end. The material is Grade S275 steel. Design the girder using the simplified method for web first.

5.4.1 Loads, shears and moments

The factored loads are:

Concentrated loads $= (1.4 \times 450) + (1.6 \times 300) = 1110\,\text{kN}$
Distributed load $= (1.4 \times 20) + (1.6 \times 10) = 44\,\text{kN/m}$

The loads and reactions are shown in Figure 5.15(a) and the shear force diagram in Figure 5.15(b). The moments are:

$M_\text{C} = (1374 \times 4) - (44 \times 4 \times 2) = 5144\,\text{kNm}$
$M_\text{E} = (1374 \times 6) - (1110 \times 2) - (44 \times 6 \times 3) = 5232\,\text{kNm}$

The bending moment diagram is shown in Figure 5.15(c).

5.4.2 Girder section for moment

(1) Design for girder depth span/10

Take the overall depth of the girder as 1200 mm and assume that the flange plates are over 40 mm thick. Then the design strength from BS 5950: Part 1, Table 9 for plates is $p_\text{y} = 255\,\text{N/mm}^2$.

The flanges resist all the moment by a couple with lever arm of, say, 1140 mm, as shown in Figure 5.16(a). The flange area is:

$$= \frac{5232 \times 10^6}{1140 \times 255} = 17\,998\,\text{mm}^2$$

Make the flange plates $450 \times 45\,\text{mm}^2$, giving an area of $20\,250\,\text{mm}^2$. The girder section with web plate 10 mm thick is shown in Figure 5.16(b).

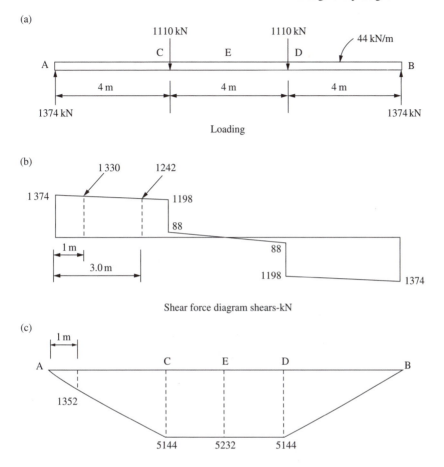

(a)

1110 kN   1110 kN

C   E   D   44 kN/m

A   B

4 m   4 m   4 m

1374 kN   1374 kN

Loading

(b)

1330   1242

1374   1198

88

1 m   88

3.0 m   88

1198   1374

Shear force diagram shears-kN

(c)

1 m

A   C   E   D   B

1352

5144   5232   5144

Bending moment diagram moments-kNm

Figure 5.15 Load, shear and moment diagrams

The flange projection b is 220 mm and the ratio:

$b/T = 220/45 = 4.89$.

Referring to Table 11 of the code, the ratio:

$$\frac{b}{T} = 4.89 \le 8\varepsilon = 8 \left(\frac{275}{255}\right)^{0.5} = 8.31$$

The flanges are Class 1 plastic and the area of cross section is $51\,600\,\text{mm}^2$.

(2) Design using the optimum depth chart

Redesign the girder using the optimum depth chart shown in Figure 5.10. Assume that the flange plates are between 16 and 40 mm thick. Then the

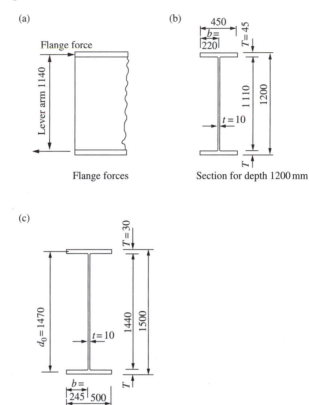

Section for depth 1200 mm

Section at optimum depth

Figure 5.16 Plate girder sections

design strength from Table 9 of the code is:

$p_y = 265 \, \text{N/mm}^2$
Plastic modulus $S_x = 5232 \times 10^1/265 = 19.74 \times 10^3 \, \text{cm}^3$
Using curve $d_0/t = 150$, the optimum depth $d_0 = 1450 \, \text{mm}$
Make the depth 1500 mm:

$$\text{Flange area} = \frac{5232 \times 10^6}{1500 \times 265} = 13\,162 \, \text{mm}^2$$

Provide flanges $500 \times 30 \, \text{mm}^2$ giving an area of $15\,000 \, \text{mm}^2$. The girder section with web plate 10 mm thick is shown in Figure 5.16(c). Note that the actual d_0/t ratio is 144.

The flange projection b is 245 mm and the ratio $b/T = 245/30 = 8.17$. Referring to the limits in Table 11 in the code, the flanges are Class 2 compact. The area of cross-section is $44\,400 \, \text{mm}^2$. The saving in material compared with the first design is 13.9 per cent.

The design will be based on a depth of 1200 mm because of headroom restriction.

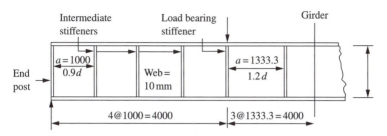

Figure 5.17 Stiffener arrangement

5.4.3 Design of web (no tension field action, $V_{cr} > F_v$)

(1) Minimum thickness of web (Section 4.4.2 of BS 5950)

An arrangement for the stiffeners is set out in Figure 5.17. The design strength of the web $p_y = 275 \, \text{kN/mm}^2$ from Table 9 of BS 5950: Part 1 for plate less than 16 mm thick. The minimum thickness is the greater of:

(1) *Serviceability.* Stiffener spacing $a >$ depth d in the centre of the girder. Web thickness $t > 1110/250 = 4.4 \, \text{mm}$.
(2) To prevent the flange buckling into the web:
    Stiffener spacing $a < 1.5$ depth d:

$$\text{Web thickness } t \geq \frac{1110}{250} \left(\frac{275}{455} \right)^{0.5} = 3.45 \, \text{mm}$$

(2) Buckling resistance of web (Section 4.4.5.2 of BS 5950)

Try a 10 mm thick web plate. The buckling resistance is checked for the maximum shear in the end panel:

Web depth/thickness ratio $d/t = 111$

Stiffener spacing/web depth ratio

$a/d = 100/1110 = 0.9$

From Table 21 in the code the shear buckling strength:

$q_w = 143 \, \text{N/mm}^2$

Shear buckling resistance:

$$V_b = V_w = 143 \times 10 \times 1110/10^3 = 1587.3 \, \text{kN}$$

Critical shear buckling resistance:

$$P_v = 0.6 P_y A_v = 0.6 \times 275 \times 1110 \times 10 \times 10^{-3} = 1831 \, \text{kN}$$

Since $P_v > V_w > 0.72 P_v$
$V_{cr} = (9V_w - 2P_v)/7 = 1517.2\,kN$
Factored applied shear $F_v = 1374\,kN < 1517.2\,kN$

The stiffener arrangement and web thickness are satisfactory. Since the critical shear buckling resistance V_{cr} of the stiffened web is sufficient to resist the applied shear force F_v, tension field action is not developed in the web. The design of the intermediate, load carrying and bearing stiffeners, and end-post is therefore greatly simplified as given below.

5.4.4 Intermediate stiffeners

(1) Trial size and outstand (Section 4.5.1.2 of BS 5950)

Try stiffeners composed of 2 No. $60 \times 8\,mm^2$ flats:

Design strength $p_y = 275\,kN/mm^2$ (Table 9)
Factor $\varepsilon = 1.0$
Outstand $60 < 13 \times 8 = 104\,mm$.

(2) Minimum stiffness (Section 4.4.6.4 of BS 5950)

The intermediate stiffener is shown in Figure 5.18. The moment of inertia about the centre of the web is:

$$I_s = 8 \times 130^3/12 = 1.464 \times 10^6\,mm^4$$
$$> \frac{1.5 \times 1110^3 \times 8^3}{1000^2} = 1.05 \times 10^6\,mm^4.$$

When the spacing $a = 1\,000\,mm < \sqrt{2}(1100) = 1569.5\,mm$.

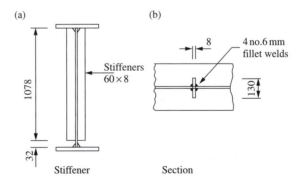

Figure 5.18 Intermediate stiffener

Note that t, the minimum required web thickness for spacing $a = 1000\,\text{mm}$ using tension field action, is 8 mm (see Section 5.5 below). The stiffener is satisfactory with respect to stiffness. In a conservative design, $t = 10\,\text{mm}$.

$$I_s \geq 1.5 \times 1110 \times 10^3/1000^2 = 2.05 \times 10^6\,\text{mm}^4$$

Stiffeners $70 \times 8\,\text{mm}^2$ are then required.

(3) Connection to web (Section 4.4.6.7 of BS 5950)

Shear between each flat and web $= 10^2/8 \times 60 = 0.208\,\text{kN/mm}$ on two welds
Use 6 mm fillet weld, strength 0.924 kN/mm.
   Four continuous fillet welds are provided.

5.4.5 Load carrying and bearing stiffener

(1) Trial size and outstand

Try stiffeners composed of 2 No. $150 \times 15\,\text{mm}^2$ plates as shown in Figure 5.19:

Outstand $150 < 13 \times 15 = 195\,\text{mm}$

The stiffener is fully effective in resisting load.

(2) Bearing check (Section 4.5.2.2 of BS 5950)

The area in bearing at the top of the stiffener is shown in Figure 5.19(b). The stiffeners have been cut back 15 mm to clear the web to flange welds:

   Design strength of stiffener $P_{ys} = 275\,\text{N/mm}^2$
   $A_{s.net} = 2 \times 15 \times 135 = 4050\,\text{mm}^2$
   $P_s = 4050 \times 275 \times 10^{-3} = 1113.8\,\text{kN} > 1110\,\text{kN}$

   The bearing capacity P_s from the stiffener itself is already sufficient, no needs to include the bearing capacity P_{bw} from the unstiffened web.

(3) Buckling check (Section 4.5.3.3 of BS 5950)

The stiffener area at the centre of the girder acting as a strut is shown in Figure 5.19(c). The stiffener properties are calculated from the dimensions shown:

   $A = (2 \times 150 \times 15) + (300 \times 10) = 7500\,\text{mm}^2$
   $I_x = 15 \times 310^3/12 = 37.23 \times 10^6\,\text{mm}^4$
   $R_x = (37.23 \times 106/8500)^{0.5} = 66.1\,\text{mm}$

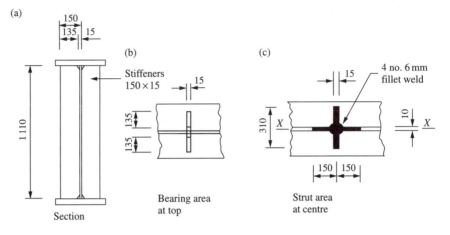

Figure 5.19 Load carrying and bearing stiffener

Assume that the flange is restrained against lateral movement and against rotation in the plane of the stiffeners:

Slenderness $\lambda = 0.7 \times 1110/66.1 = 11.8$

Design strength $= 275\,N/mm^2$ (No reduction necessary for welded
stiffener)

Compressive strength $p_c = 275\,N/mm^2$ (Table 24 for strut curve c)

Buckling resistance:

$$P_x = 275 \times 7500/10^3 = 2062.5\,kN$$

The size selected is satisfactory.

(4) Connection to web

Shear between each flat and web:

$$= \frac{10^2}{8 \times 150} + \frac{1110}{2 \times 1110} = 0.583\,kN/mm \text{ on two welds.}$$

Use 6-mm continuous fillet weld, strength is $0.924\,kN/mm$. Four fillet welds are provided. Note that the bearing area required controls the stiffener size.

5.4.6 End-post

(1) Trial size and outstand

The trial size for the end-post consisting of a single plate $450 \times 15\,mm^2$ is shown in Figure 5.20(a). The end-post is also designed as a load carrying and

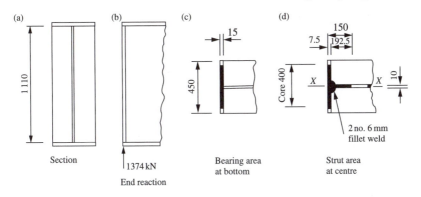

Figure 5.20 End-post

bearing stiffener because no tension field action is necessary in the end panel, no anchorage and hence no anchor force is developed.

$$\text{Outstand} = 220\,\text{mm} > 13 \times 15 = 195\,\text{mm}$$
$$< 19 \times 15 = 285\,\text{mm}$$

Base design on a stiffener core 400 mm × 15 mm
Design strength $= 275\,\text{N/mm}^2$ (Table 9)

(2) Bearing check

The bearing area is shown in Figure 5.20(c):

$$A_{\text{s.net}} = 15 \times 400 = 6000\,\text{mm}^2$$
$$P_{\text{s}} = 6000 \times 275 \times 10^{-3} = 1650\,\text{kN} > 1374\,\text{kN (satisfactory)}$$

(3) Buckling check

The area at the centre line acting as a strut is shown in Figure 5.20(d):

$$A = (400 \times 15) + (142.5 \times 10) = 7425\,\text{mm}^2$$
$$I_x = 15 \times 400^3/12 = 80 \times 10^6\,\text{mm}^4$$
$$r_x = (80.0 \times 10^6/7925)^{0.5} = 100.4$$
$$\lambda = 0.7 \times 1110/100.4 = 7.7$$
$$p_c = 275\,\text{N/mm}^2 \text{ (Table 24 for strut curve } c)$$
$$P_x = 275 \times 7425/10^3 = 2041.9\,\text{kN}$$
$$\text{Load carried} = 1374\,\text{kN}$$

The size is satisfactory.

(4) Connection to web

Shear between end plate and web:

$$= \frac{1374}{2 \times 1110} = 0.62 \text{ kN per weld}$$

Provide 6-mm continuous fillet, weld strength 0.924 kN/mm. Two lengths of weld are provided.

5.4.7 Flange to web weld

See Figure 5.16(b) for the girder dimension:

$$I_x = (450 \times 1200^3 - 440 \times 1110^3)/12 = 14.65 \times 10^9 \text{ mm}^4$$

Horizontal shear per weld (see Section 5.3.9):

$$= \frac{1374 \times 450 \times 45 \times 577.5}{14.65 \times 10^9 \times 2} = 0.548 \text{ kN/mm}$$

Provide 6-mm continuous fillet, weld strength 0.924 kN/mm.

5.4.8 Design drawing

A design drawing of the girder is shown in Figure 5.21.

5.5 Design utilizing tension field action ($V_b = V_w + V_f$)

Redesign the web, stiffeners and end post for the girder in Section 5.4 using the more exact method for web (i.e. utilizing tension field action in the web).

5.5.1 Design of the web

Try an 8-mm thick web with the stiffeners spaced at 1000 mm in the end 4 m of the girder, as shown in Figure 5.22. The web design is set out in Section 4.4.5.3 of BS 5950:

$$d/t = 1110/8 = 138.75$$
$$a/d = 1000/1110 = 0.9$$

The shear buckling strength from Table 21 in the code:

$$q_w = 118 \text{ N/mm}^2$$

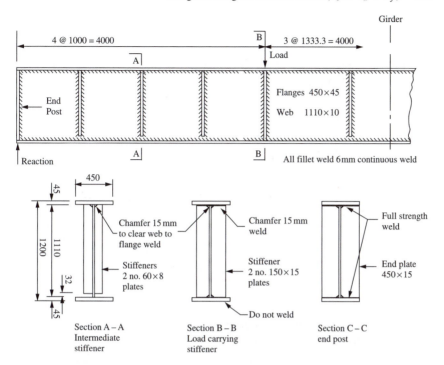

Figure 5.21 Design without utilizing tension field action

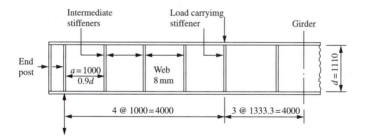

Figure 5.22 Stiffener arrangement

The shear buckling resistance of the stiffened panel is:

$$V_b = V_w = 118 \times 1110 \times 8/10^3 = 1047.8\,\text{kN}$$

This is less than the applied shear of 1374 kN. The contribution to shear buckling resistance from the flanges is now necessary; hence use the more exact method.

$$V_b = V_w + V_f \quad \text{but} \quad V_b \le P_v$$

The plastic moment capacity of the flange where the design strength of the flange $p_{yf} = 255\,\text{N/mm}^2$:

$$M_{pf} = \frac{255 \times 450 \times 45^2}{4 \times 10^6} = 58.1\,\text{kNm}$$

The plastic moment capacity of the web where the design strength of the web $p_{yw} = 275\,\text{N/mm}^2$:

$$M_{pw} = \frac{275 \times 8 \times 1110^2}{4 \times 10^6} = 677.6\,\text{kNm}$$

The maximum moment in the end panel is 1352 kNm (see Figure 5.15(c)). The mean longitudinal stress in the flange due to moment:

$$f_f = \frac{1352 \times 10^6}{1155 \times 45 \times 450} = 57.8\,\text{N/mm}^2$$
$$P_v = 0.6 P_y A_v = 0.6 \times 275 \times 8 \times 1110 \times 10^{-3} = 1465.2\,\text{kN}$$

Since the flanges are not fully stressed ($f_f < P_{yf}$), $V_b = V_w + V_f$ but $V_b \le P_v$. The flange-dependent shear buckling resistance:

$$
\begin{aligned}
V_f &= \frac{P_v\,(d/a)\left[1 - \left(f_f/p_{yf}\right)^2\right]}{1 + 0.15\left(M_{pw}/M_{pf}\right)} \\
&= \frac{1465.2(1100/1000)\left[1 - (57.8/255)^2\right]}{1 + 0.15(677.6/58.1)} = 556.1\,\text{kN}
\end{aligned}
$$

The total shear buckling resistance:

$$V_b = 1047.8 + 556.1 = 1603.9 \le 1465.2\,\text{kN}$$

This exceeds the applied shear of 1374 kN (hence, satisfactory).

Check the web in the panel at 3.0 m from the support:

Applied shear $= 1242\,\text{kN}$

$$f_f = \frac{3924 \times 10^6}{1155 \times 45 \times 450} = 167.8\,\text{N/mm}^2$$

$$\begin{aligned} V_f &= \frac{P_v\,(d/a)\left[1 - \left(f_f/p_{yf}\right)^2\right]}{1 + 0.15\left(M_{pw}/M_{pf}\right)} \\ &= \frac{1465.2(1100/1000)\left[1 - (167.8/255)^2\right]}{1 + 0.15(677.6/58.1)} = 330.9\,\text{kN} \end{aligned}$$

Total shear buckling resistance:

$$V_b = 1047.8 + 330.9 = 1378.7 \le 1465.2\,\text{kN}$$

The girder is satisfactory for the stiffener arrangement assumed.

5.5.2 Intermediate stiffeners

(1) Minimum stiffness

Try stiffeners composed of two No. $80 \times 8\,\text{mm}^2$ flats (see Section 5.4.4 above). The outstand is satisfactory. The stiffener is shown in Figure 5.23:

$$I_s = 8 \times 168^3/12 = 3.161 \times 10^6\,\text{mm}^4 > 1.05 \times 10^6\,\text{mm}^4$$

(2) Buckling check

Maximum shear adjacent to the stiffener at 1.0 m from support (see Figure 5.15(b)):

$$V = 1330\,\text{kN}$$

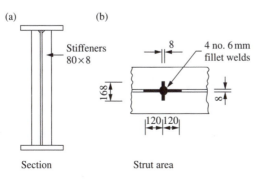

Figure 5.23 Intermediate stiffener

The critical shear buckling resistance of web:

$$V_w = 1047.8 \, kN$$

$$P_v = 0.6 P_y A_v = 0.6 \times 275 \times 1110 \times 8 \times 10^{-3} = 1465.2 \, kN$$

Since $V_w \leq 0.72 P_v \quad V_{cr} = (V_w/0.9)^2 / P_v = 924.9 \, kN$

Stiffener force $F_q = V - V_{cr} = 1330 - 924.9 = 405.1 \, kN$

The stiffener properties are:

$$A = (160 \times 8) + (240 \times 8) = 3200 \, mm^2$$

$$r_x = (3.161 \times 10^6 / 3200)^{0.5} = 31.43 \, mm$$

$$\lambda = 0.7 \times 1110/31.43 = 24.7$$

$$p_c = 254 \, N/mm^2 \text{ from Table 24(curve } c) \text{ for } p_y = 255 \, N/mm^2$$

Buckling resistance:

$$P_q = 254 \times 3200/10^3 = 812.8 \, kN > F_q$$

The stiffener is satisfactory. (Note: these stiffeners are to extend from flange to flange, not permitted to terminate clear of the tension flange in this case, see clause 4.4.6.7 of the code).

(3) Connection to web

Provide 6-mm continuous fillet weld.

5.5.3 Load carrying and bearing stiffener

Try stiffeners composed of 2 Nos. $150 \times 20 \, mm^2$ plates as shown in Figure 5.24. The stiffener outstand will be satisfactory and the bearing check will also be adequate (refer to Section 5.4.5 earlier).

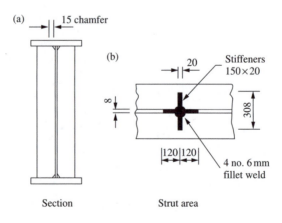

Figure 5.24 Load carrying and bearing stiffener

Only the buckling check is carried out here:

$$A = (2 \times 150 \times 20) + (240 \times 8) = 7920\,\text{mm}^2$$
$$I_x = 20 \times 308^3/12 = 48.69 \times 10^6\,\text{mm}^4$$
$$r_x = (48.69 \times 10^6/7920)^{0.5} = 78.4\,\text{mm}$$
$$\lambda = 0.7 \times 1110/78.4 = 9.91$$
$$p_y = 275\,\text{N/mm}^2 \text{ (Table 9)}$$
$$p_c = 275\,\text{N/mm}^2 \text{ (Table 24 for curve } c)$$
$$P_x = 275 \times 7920/10^3 = 2178\,\text{kN}$$

Combined external transverse shear force $F_x = 1110 + 1198 - 924.9 = 1383.1\,\text{kN} < P_x$

The size selected for the stiffener is satisfactory. Provide 6-mm continuous fillet weld between stiffeners and web.

5.5.4 End-post

The design will be made using twin stiffener end post as shown in Figure 5.25.

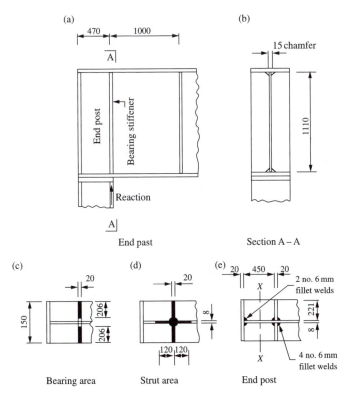

Figure 5.25 End-post

(1) Bearing check

The reaction is carried on the inner stiffener. The stiffener ends are chamfered to clear the web to flange welds and the bearing area is shown in Figure 5.25(c).

$p_{ys} = 265\,\text{N/mm}^2$ (Table 9)

$A_{s.net} = 2 \times 20 \times 206 = 8240\,\text{mm}^2$

$P_s = 265 \times 8240 = 2183.6\,\text{kN} > 1374\,\text{kN}$

Note that outstand $= 221\,\text{mm} < 13 \times 20 \left(\dfrac{275}{265}\right)^{0.5} = 264.9\,\text{mm}$

The full area of the stiffener is effective and the stiffener is satisfactory for bearing.

(2) Buckling check

The area at the centre line of the bearing stiffener acting as a strut is shown in Figure 5.25(d):

$A = (2 \times 20 \times 221) + (2 \times 120 \times 8) = 10760\,mm^2$

$I_x = 20 \times 450^3/12 = 151.87 \times 10^6\,\text{mm}^4$

$r_x = (151.87 \times 10^6/10760)^{0.5} = 118.8\,\text{mm}$

$\lambda = 0.7 \times 1110/118.8 = 6.54$

$p_c = 265\,\text{N/mm}^2$ (Table 27 curve c)

$P_x = 265 \times 10760/10^3 = 2851\,\text{kN}$

This exceeds the reaction of 1374 kN. Therefore satisfactory.

(3) Shear and moment from the tension field anchor force

Refer to Section 4.4.5.4 and Annex H.4 of BS 5950.

$d/t = 138.75$ and $a/d = 0.9$

From Table 21, shear buckling strength $q_w = 118\,\text{N/mm}^2$.

The simple shear buckling resistance $V_w = 118 \times 1110 \times 8/10^3 = 1047.8\,\text{kN}$

Applied shear force in the end panel: $F_v = 1374\,\text{kN}$

Since $F_v > V_w$, the web is fully loaded in shear and the tension field longitudinal anchor force H_q cannot be reduced. This anchor force is:

$H_q = 0.5\,dtp_y(1 - V_{cr}/P_v)^{0.5}$

$= 0.5 \times 1100 \times 8 \times 275 \times 10^{-3}\{1 - (924.9/1465.2)\}^{0.5}$

$= 734.8\,\text{kN}$

Shear from the tension field force:

$$R_{tf} = 0.75 \, H_q = 551.1 \, \text{kN}$$

Moment in the end-post:

$$M_{tf} = \frac{0.15 \times 734.8 \times 1110}{10^3} = 121.2 \, \text{kNm}$$

(4) Shear capacity of end-post

The end-post is shown in Figure 5.25(e). The web $450 \times 8 \, \text{mm}^2$ resists shear (see Table 11 of code for limiting proportions for webs):

$d/t = 450/8 = 56.25 < 80\varepsilon$

The section is Class 1 Plastic

Shear capacity, $P_v = 0.6 \times 275 \times 450 \times 8/10^3 = 594 \, \text{kN}$

The end-post is satisfactory with respect to shear.

(5) Moment capacity

Referring to Figure 5.25(e), the flange proportions are:

$b/T = 221/20 = 11.05$

Design strength $p_y = 265 \, \text{N/mm}^2$ (Table 9 in the code).
From Table 11, $b/T < 13\varepsilon = 13.2$. The flanges are Class 3 semi-compact. Moment capacity check at the centre of the girder:

$M_{cx} = 265 \times 450 \times 20 \times 470/10^6 = 1120.9 \, \text{kNm}.$

This exceeds the moment from tension field action of 121.1 kNm.

(6) Additional stiffener force due to moment

The moment M_{tf} induces additional compressive force F_{tf} in the inner stiffener which must be checked for (see Section H.4.3 of the code).

$F_{tf} = M_{tf}/a_e$ where a_e is the centre-to-centre of the twin stiffeners.
$F_{tf} = 121.1 \times 10^3/470 = 257.7 \, \text{kN}$

The inner twin stiffener is still satisfactory for bearing and buckling with this additional force (see Section 6.5.4 (1) and (2)).

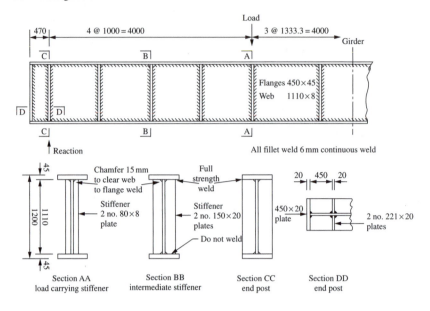

Figure 5.26 Design utilizing tension field action

(7) Weld sizes

The four fillet welds shown in Figure 5.25(e) to connect the bearing stiffener to the web is designed first. The welds must support the reaction and the beam shear from the end-post:

$$\text{End-post } I_x = (450 \times 490^3 - 442 \times 450^3)/12$$
$$= 1055 \times 10^6 \text{ mm}^4$$
$$\text{Weld force} = \frac{1374}{4 \times 1110} + \frac{551.1 \times 2 \times 221 \times 20 \times 234}{4 \times 1055 \times 10^6}$$
$$= 0.31 + 0.27$$
$$= 0.58 \text{ kN/mm}$$

Provide 6 mm continuous fillet weld; strength 0.924 kN/mm.

This size of weld will be satisfactory for the welds between the end-plate and web.

5.5.5 Design drawing

A drawing of the girder designed using the more exact method for the web and utilizing tension field action is shown in Figure 5.26.

Problems

5.1 A welded plate girder fabricated from Grade S275 steel is proportioned as shown in Figure 5.27. It spans 15.0 m between centres of brackets and supports a 254 × 254 UC 107 column at mid-span. The loading is shown

in the figure. The compression flange is effectively restrained over the span
and intermediate stiffeners are provided at 1.875 m centres between supports
and the centre load. Assuming that the plate girder and fire-protection casing
weigh 20 kN/m, carry out the following design work:

(1) Check the adequacy of the section with respect to bending, shear and
    deflection.
(2) Design a suitable load carrying and bearing stiffener for the supports
    and concentrated load positions.
(3) Determine the weld size required at the point of maximum shear.

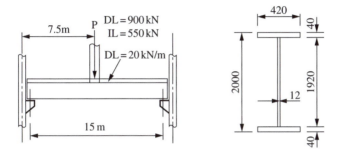

Figure 5.27

5.2 A welded plate girder of Grade S275 steel carries two concentrated loads
transmitted from 254×254 UB 107 columns at the third points. The columns
rest on the top flange and the loads are each 400 kN dead load and 250 kN
imposed load, respectively. The plate girder is 12 m span and is simply
supported at its ends. The compression flange has adequate lateral restraint
at the points of concentrated loads and at the supports. Assume that the
weight of the girder is 4 kN/m and that the girder is supported on brackets
at each end.

(1) Assuming that the depth limit is 1400 mm for the plate girder, design
    the girder section.
(2) Design the intermediate, load carrying and bearing stiffeners.
(3) Design the web-to-flange weld.
(4) Sketch the arrangement and details of the plate girder.

5.3 The framing plans for a four-storey building are shown in Figure 5.28. The
front elevation is to have a plate girder at first-floor level to carry wall and
floors and give clear access between columns B and C. The plate girder is
simply supported with a shear connection between the girder end plates and
the column flanges. Columns B and C are 305×305 UC 158. The loading
from floor, roof and wall is as follows:

Dead loads:
Front wall between B and C
(includes glazing and columns) $= 0.7 \, kN/m^2$
Floors of r.c. slab:
(includes screed, finish, ceilings, etc.) $= 6.0 \, kN/m^2$
Roof of r.c. slab:
(includes screed, finish, ceilings, etc.) $= 4.0 \, kN/m^2$

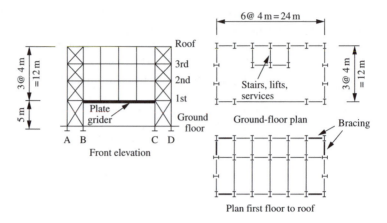

Figure 5.28 Framing plan for a three-storey building

Imposed loads:
Roof $= 1.5\,\text{kN/m}^2$
Floors $= 2.5\,\text{kN/m}^2$

(1) Calculate the loads on the girder.
(2) Design the plate girder and show all design information on a sketch.

6
Tension members

6.1 Uses, types and design considerations

6.1.1 Uses and types

A tension member transmits a direct axial pull between two points in a structural frame. A rope supporting a load or cables in a suspension bridge are obvious examples. In building frames, tension members occur as:

(1) tension chords and internal ties in trusses;
(2) tension bracing members;
(3) hangers supporting floor beams.

Examples of these members are shown in Figure 6.1.
   The main sections used for tension members are:

(1) open sections such as angles, channels, tees, joists, universal beams and columns;
(2) closed sections. Circular, square and rectangular hollow sections;
(3) compound and built-up sections. Double angles and double channels are common compound sections used in trusses. Built-up sections are used in bridge trusses.

Round bars, flats and cables can also be used for tension members where there is no reversal of load. These elements as well as single angles are used in cross bracing, where the tension diagonal only is effective in carrying a load, as shown in Figure 6.1(d). Common tension member sections are shown in Figure 6.2.

6.1.2 Design considerations

Theoretically, the tension member is the most efficient structural element, but its efficiency may be seriously affected by the following factors:

(1) The end connections. For example, bolt holes reduce the member section.
(2) The member may be subject to reversal of load, in which case it is liable to buckle because a tension member is more slender than a compression member.

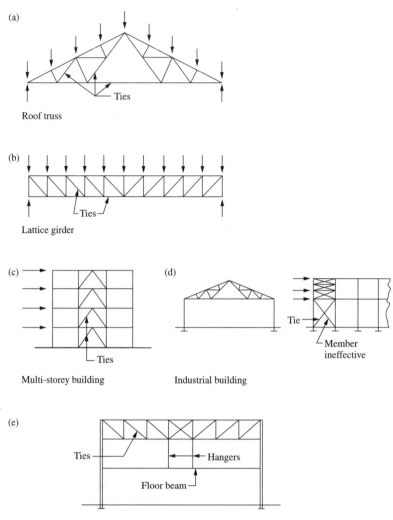

(a)

Ties

Roof truss

(b)

Ties

Lattice girder

(c)   (d)

Ties   Tie

Multi-storey building   Industrial building   Member
                                                                ineffective

(e)

Ties ——   —— Hangers

Floor beam ——

Hangers supporting floor beam

Figure 6.1 Tension members in buildings

(3) Many tension members must also resist moment as well as axial load. The moment is due to eccentricity in the end connections or to lateral load on the member.

6.2 End connections

Some common end connections for tension members are shown in Figures 6.3(a) and (b). Comments on the various types are:

(1) Bolt or threaded bar. The strength is determined by the tensile area at the threads.

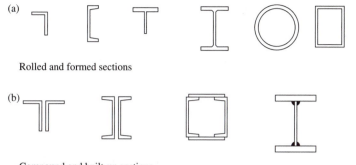

(a)

Rolled and formed sections

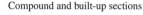

(b)

Compound and built-up sections

Figure 6.2 Tension member sections

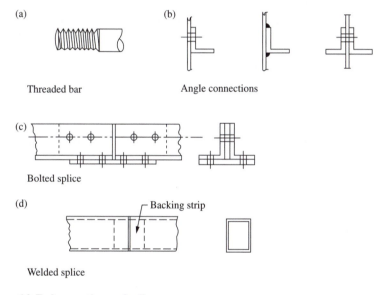

(a)

Threaded bar

(b)

Angle connections

(c)

Bolted splice

(d)   Backing strip

Welded splice

Figure 6.3 End connections and splices

(2) Single angle connected through one leg. The outstanding leg is not fully
     effective, and if bolts are used the connected leg is also weakened by the
     bolt hole.

Full-strength joints can be made by welding. Examples occur in lattice girders
made from hollow sections. However, for ease of erection, most site joints are
bolted, and welding is normally confined to shop joints.

    Site splices are needed to connect together large trusses that have been
fabricated in sections for convenience in transport. Shop splices are needed in
long members or where the member section changes. Examples of bolted and
welded splices in tension members are shown in Figures 6.2(c) and (d).

6.3 Structural behaviour of tension members

6.3.1 Direct tension

The tension member behaves in the same way as a tensile test specimen. In the elastic region:

Tensile stress $f_t = P/A$,
Elongation $\delta = PL/AE$

where P is the load on the member, A the area of cross section and L the length.

6.3.2 Tension and moment: elastic analysis

(1) Moment about one axis

Consider the I-section shown in Figure 6.4(b), which has two axes of symmetry. If the section is subjected to an axial tension P and moment M_x about the x–x axis, the stresses are:

Direct tensile stress $f_t = P/A$,
Tensile bending stress $f_{bx} = M_x/Z_x$,
Maximum tensile stress $f_{max} = f_t + f_{bx}$,

where Z_x is the elastic modulus for the x–x axis.
The stress diagrams ace shown in Figure 6.4(b). Define the allowable stresses:

p_t—direct tension,

p_{bt}—tension due to bending.

Then the interaction expression

$$\frac{f_t}{P_t} + \frac{f_{bt}}{P_{bt}} \leq 1$$

gives permissible combinations of stresses. This is shown graphically in Figure 6.4(c). A section with one axis of symmetry may be treated similarly.

(2) Moment about two axes

If the section is subjected to axial tension P and moments M_x and M_x about the x–x and y–y axes, respectively, the individual stresses and maximum stress are:

Direct tensile stress $f_t = P/A$,
Tensile bending stress x–x axis $f_{bx} = M_x/Z_x$,
Tensile bending stress y–y axis $f_{by} = M_y/Z_y$,
Maximum stress $f_{max} = f_t + f_{bx} + f_{by}$
Z_y = elastic modulus for the y–y axis.

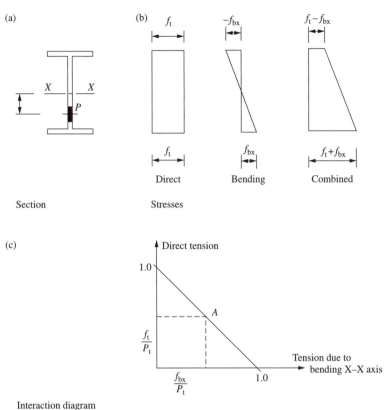

(a)

(b)

Section

Stresses

Direct   Bending   Combined

(c)

Interaction diagram

Figure 6.4 Elastic analysis: tension and moment about one axis

These stresses are shown in Figures 6.5(b)–(d).
The interaction expression to give permissible combinations of stresses is:

$$\frac{f_t}{P_t} + \frac{f_{bx}}{P_{bt}} + \frac{f_{by}}{P_{bt}} \le 1$$

This may be represented graphically by the plane in Figure 6.5(e).

Sections with one axis of symmetry or with no axis of symmetry which are free to bend about the principal axes can be treated similarly.

6.3.3 Tension and moment: plastic analysis

(1) Moment about one axis

For a section with two axes of symmetry (as shown in Figure 6.6(a)), the moment is resisted by two equal areas extending inwards from the extreme fibres. The central core resists the axial tension. The stress distribution is shown in Figure 6.6(b) for the case where the tension area lies in the web. At higher loads, the area needed to support tension spreads to the flanges, as shown in Figure 6.6(c).

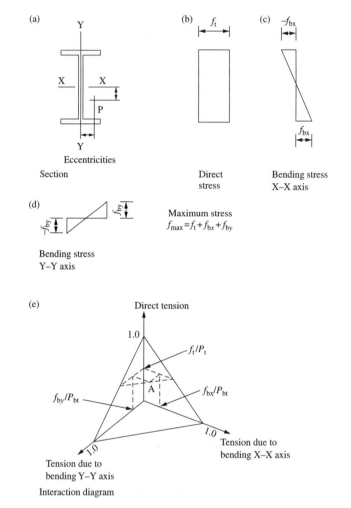

(a)

Y

Eccentricities

Section

(b) f_t

Direct
stress

(c) $-f_{bx}$

f_{bx}

Bending stress
X–X axis

(d)

$-f_{by}$

f_{by}

Bending stress
Y–Y axis

Maximum stress
$f_{max} = f_t + f_{bx} + f_{by}$

(e)

Direct tension

1.0

f_t/P_t

f_{by}/P_{bt}

A

f_{bx}/P_{bt}

1.0

Tension due to
bending X–X axis

1.0

Tension due to
bending Y–Y axis

Interaction diagram

Figure 6.5 Elastic analysis: tension and moments about two axes

For design strength p_y, the maximum tension the section can support is:

$$P_t = p_y A.$$

If moment only is applied, the section can resist:

Plastic moment $M_{cx} = p_y S_x$
Elastic moment $M_{Ex} = p_y Z_x$
where S_x denotes plastic modulus and Z_x the elastic modulus.
    For values F of tension less than P_t if the tension area is in the web (as shown in Figure 6.6(a), the length a of web supporting F is:

$$a = F/(p_y t).$$

where t is the web thickness.

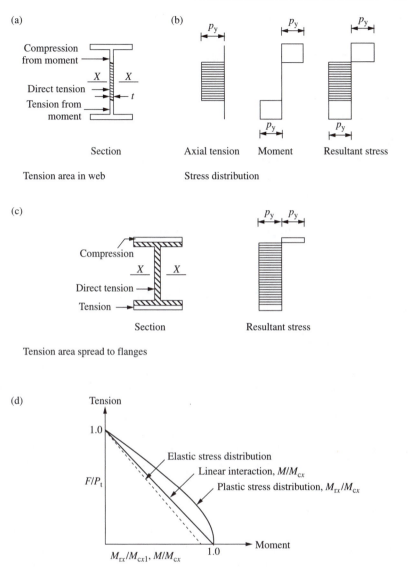

(a)

Compression
from moment

Direct tension

Tension from
moment

X X

t

Section

Tension area in web

(b)

p_y p_y p_y

p_y p_y

Axial tension   Moment   Resultant stress

Stress distribution

(c)

Compression

Direct tension

Tension

X X

Section

Tension area spread to flanges

p_y p_y

Resultant stress

(d)

Tension

1.0

F/P_t

Elastic stress distribution

Linear interaction, M/M_{cx}

Plastic stress distribution, M_{rx}/M_{cx}

Moment

1.0

M_{rx}/M_{cx1}, M/M_{cx}

Interaction diagram

Figure 6.6 Plastic analysis: tension and moment about one axis

The reduced moment capacity in the presence of axial load is:

$$M_{rx} = (S_x - ta^2/4)\,p_y.$$

A more complicated formula is needed for the case where the tension area enters the flanges, as shown in Figure 6.6(b). The curve of F/P_t against M_{rx}/M_{cx}, for an I-section bent about the x–x axis is convex (see Figure 6.6(d)), but a conservative design results if the straight line joining the end points is adopted. This gives the linear interaction expression:

$$\frac{F}{P_t} + \frac{M_x}{M_{cx}} = 1$$

where M_x is the applied moment.

The elastic curve is also shown. The strength gain due to plasticity is the area between the two curves.

In calculating the reduced moment capacity, it is convenient to use a reduced plastic modulus. This was given for the case above by:

$$S_{rx} = S_x - ta^2/4.$$

If the average stress on the whole section of area A:

$$f = F/A,$$

then the formula for reduced plastic modulus can be written after substituting for a as:

$$S_{rx} = S_x - n^2 A^2/4t$$

where $n = f/p_y$.

The expression is more complicated when the tension area spreads into the flanges.

These are the formulae given in *Steelwork Design*, Guide to BS 5980: Part 1: Volume 1, to calculate the reduced plastic modulus. The change value of n is given to indicate when the tension area enters the flanges. Note that the reduced plastic modulus is not required if the linear interaction expression is adopted. The analysis for sections with one axis of symmetry is more complicated.

(2) *Moment about two axes*

Solutions can be found for sections subject to axial tension and moments about both axes at full plasticity. I-sections with two axes of symmetry have been found to give a convex failure surface, as shown in Figure 6.7. This interaction surface is constructed in terms of:

$$F/P_t, \quad M_{rx}/M_{cx}, \quad M_{ry}/M_{cy}$$

Where F is the axial tension, P_t the tension capacity, M_{cx} the moment capacity for the x–x axis in the absence of axial load, M_{rx} the reduced moment capacity for the x–x axis in the presence of axial load, M_{cy} the moment capacity for the y–y axis in the absence of axial load, and M_{ry} the reduced moment capacity for the y–y axis in the presence of axial load.

In practice, M_{cy} is restricted with some sections. Any point A on the failure surface gives the permissible combination of axial load and moments the section can support.

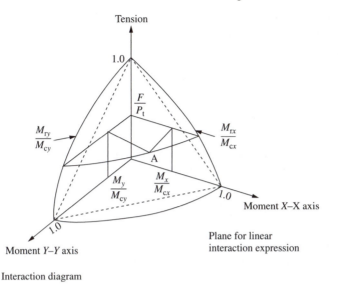

Interaction diagram

Figure 6.7 Plastic analysis: tension and moment about two axes

A plane may be drawn through the terminal points on the failure surface. This can be used to give a simplified and conservative linear interaction expression:

$$\frac{F_t}{P_t} + \frac{M_x}{M_{cx}} + \frac{M_y}{M_{cy}} = 1$$

where M_x is the applied moment about the x–x axis and M_y the applied moment about the y–y axis.

6.4 Design of tension members

6.4.1 Axially loaded tension members

The tension capacity is given in Section 4.6.1 of BS 5950: Part 1. This is:

$$P_t = A_e p_y$$

where A_e is the effective area of the section defined in Sections 3.4.3, 4.6.2 and 4.6.3 of the code.

From Section 3.4.3, the effective area of each element of a member is given by:

$A_e = K_e \times$ net area where holes occur $\leq$ gross area
$K_e = 1.2$ for Grade S275 and 1.1 for S355 steel
(Net area = gross area less holes.)

Tests show that holes do not reduce the capacity of a member in tension provided that the ratio of net area to gross area is greater than the ratio of yield strength to ultimate strength.

6.4.2 Simple tension members

(1) Single angles, channels or T-section members
connected through one leg

These may be designed in accordance with Section 4.6.3 of the code as axially
loaded members with an effective area (see Figure 6.3(b)):

For bolted connection: $P_t = p_y(A_e - 0.5a_2)$
For welded connection: $P_t = p_y(A_g - 0.3a_2)$

where a_2 equals $(A_g - a_1)$, where A_g is the gross cross-sectional area and a_1
the gross sectional area of the connected leg.

(2) Double angles, channels or T-section members
connected through one side of a gusset

For bolted connection: $P_t = p_y(A_e - 0.25a_2)$,
For welded connection: $P_t = p_y(A_g - 0.15a_2)$.

(3) Double angles, channels or T-section members
connected to both sides of gusset plates

If these members are connected together as specified in the code, they can be
designed as axially loaded members using the net area specified in Section 3.3.2
of the code. This is the gross area minus the deduction for holes.

6.4.3 Tension members with moments

The code states in Sections 4.6.2 and 4.8.1 that moments from eccentric
end connections and other causes must be taken into account in design.
Single angles, double angles and T-sections carrying direct tension only
may be designed as axially loaded members, as set out in Section 4.6.3 of
the code.

Design of tension members with moments is covered in Section 4.8.2 of the
code. This states that the member should be checked for capacity at points of
greatest moment using the simplified interaction expression:

$$\frac{F}{A_e p_y} + \frac{M_x}{M_{cx}} + \frac{M_y}{M_{cy}} \leq 1$$

where F is the applied axial load, A_e the effective area, M_x the applied moment
about the x–x axis, M_{cx} the moment capacity about the x–x axis in the absence
of axial load, and M_y the applied moment about the y–y axis. M_{cy} the moment
capacity about the y–y axis in the absence of axial load.

The interaction expression was discussed in Section 6.3.3(2) above. (See
Section 5.4.2 for calculation of M_{cx} and M_{cy}.) For bending about one axis, the
terms for the other axis are deleted.

An alternative expression given in the code takes account of convexity of
the failure surface. This leads to greater economy in the design of plastic and
compact sections.

6.5 Design examples

6.5.1 Angle connected through one leg

Design a single angle to carry a dead load of 70 kN and an imposed load of 35 kN.

(1) Bolted connection

Factored load $= (1.4 \times 70) + (1.6 \times 35) = 154$ kN.
Try $80 \times 60 \times 7$ angle connected through the long leg, as shown in Figure 6.8(a). The bolt hole is 22 mm diameter for 20 mm diameter bolts. Design strength from Table 6 in the code $p_y = 275$ N/mm^2

$a_1 =$ net area of connected leg $= (76.5 - 22)7 = 381.5$ mm^2,

$a_2 =$ area of unconnected leg $= 56.5 \times 7 = 395.5$ mm^2,

Effective area $A_e = a_1 + a_2 = 777$ mm^2.

Tension capacity: $P_t = p_y(A_e - 0.5a_2) = 275(777 - 0.5 \times 395.5)/10^3$
$$= 159 \text{ kN.}$$

The angle is satisfactory.
   Note that the connection would require either 3 No. Grade 8.8 or 3 No. friction-grip 20 mm diameter bolts to support the load.

(2) Welded connection

Try $75 \times 50 \times 6$ L connected through the long leg (see Figure 6.8(b)):

$a_1 = 72 \times 6 = 432$ mm^2,

$a_2 = 47 \times 6 = 282$ mm^2,

$A_g = a_1 + a_2 = 714$ mm^2.

Tension capacity:

$$P_t = p_y(A_g - 0.3a_2) = 275(714 - 0.3 \times 282)/10^3 = 173 \text{ kN.}$$

The angle is satisfactory.

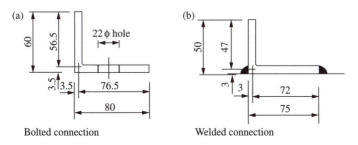

Figure 6.8 Single angle connected through one leg

6.5.2 Hanger supporting floor beams

A high-strength Grade S460 steel hanger consisting of a $203 \times 203\,\mathrm{UC}\ 46$ carries the factored loads from beams framing into it and from the floor below, as shown in Figure 6.9(a). Check the hanger at the main floor beam connection.

The design strength from Table 9 of BS 5950: Part 1 for sections less than 16 mm thick is:

$$p_y = 460\,\mathrm{N/mm^2}.$$

The net section is shown in Figure 6.9(b). For S460 steel the effective section is equal to the net section. The factor K_e from Section 3.4.3 of the code is 1.0. The connection plates are not considered.

Check the limiting proportions of the flanges using Table 6a in the code:

$$\varepsilon = (275/460)^{0.5} = 0.773,$$
$$b/t = 101.6/11 = 9.23 < 15\varepsilon = 11.72.$$

Values of b and t are shown in Figure 6.9(b).

The section is semi-compact. The moment capacity is calculated using the elastic properties. This can be calculated using first principles, and the properties are:

Location of the centroidal axis is shown.

Effective area $= 53.1\,\mathrm{cm^2}$.
Minimum value of elastic modulus $Z = 363\,\mathrm{cm^3}$.

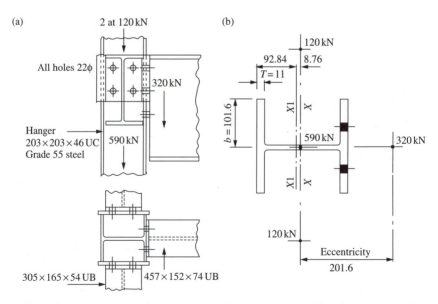

(a)

All holes 22ϕ

2 at 120 kN

320 kN

Hanger
$203 \times 203 \times 46$ UC
Grade 55 steel

590 kN

$305 \times 165 \times 54$ UB $457 \times 152 \times 74$ UB

Connection

(b)

120 kN

92.84   8.76

$T = 11$

$b = 101.6$

X1   X

590 kN   320 kN

X1   X

120 kN

Eccentricity
201.6

Hanger-net section and bolt holes

Figure 6.9 High strength hanger

The moment capacity for the major axis:

$$M_{cx} = 363 \times 460/10^3 = 166.9\,\text{kN m}.$$

The applied axial load:

$$F = (2 \times 120) + 590 + 320 = 1150\,\text{kN m}.$$

The applied moment about the x_1–x_1 axis:

$$M_x = (320 \times 0.21) + (2 \times 120 + 590)0.0088 = 74.5\,\text{kN m}.$$

Substitute into the interaction expression:

$$\frac{F}{A_e p_y} + \frac{M_x}{M_{cx}} = \frac{1150 \times 10}{53.1 \times 450} + \frac{74.5}{166.9} = 0.92 < 1.$$

The hanger is satisfactory.

Problems

6.1 A tie member in a roof truss is subjected to an ultimate tension of 1000 kN. Design this member using Grade S275 steel and an equal angle section.

6.2 A tension member in Grade S275 steel consists of 2 No. $150 \times 100 \times 8$ mm unequal angles placed back to back. At the connection, two rows of 2 No. 22 mm diameter holes are drilled through the longer legs of the angles. Determine the ultimate tensile load that can be carried by the member.

6.3 A tension member from a heavy truss is subjected to an ultimate axial load and bending moment of 2000 kN and 500 kN m, respectively. Design a suitable universal beam section in Grade S275 steel. Assume that the gross section will resist the load and moment.

6.4 A tie member in a certain steel structure is subjected to tension and biaxial bending. The ultimate tensile load was found to be 3000 kN while the ultimate moments about the major and minor axes were 160 kN m and 90 kN m, respectively. Check whether a 305×305 UC 158 in Grade S275 steel is adequate. Assume that the gross section resists the loads and moments.

7

Compression members

7.1 Types and uses

7.1.1 Types of compression members

Compression members are one of the basic structural elements, and are described by the terms 'columns', 'stanchions' or 'struts', all of which primarily resist axial load.

Columns are vertical members supporting floors, roofs and cranes in buildings. Though internal columns in buildings are essentially axially loaded and are designed as such, most columns are subjected to axial load and moment. The term 'strut' is often used to describe other compression members such as those in trusses, lattice girders or bracing. Some types of compression members are shown in Figure 7.1. Building columns will be discussed in this chapter and trusses and lattice girders are dealt with in Chapter 8.

7.1.2 Compression member sections

Compression members must resist buckling, so they tend to be stocky with square sections. The tube is the ideal shape, as will be shown below. These are in contrast to the slender and more compact tension members and deep beam sections.

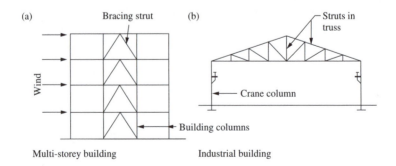

Figure 7.1 Types of compression members

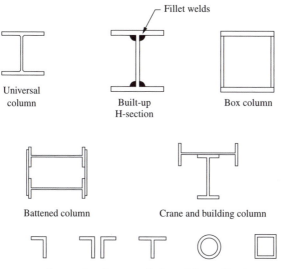

Figure 7.2 Compression member sections

Rolled, compound and built-up sections are used for columns. Universal columns are used in buildings where axial load predominates, and universal beams are often used to resist heavy moments that occur in columns in industrial buildings. Single angles, double angles, tees, channels and structural hollow sections are the common sections used for struts in trusses, lattice girders and bracing. Compression member sections are shown in Figure 7.2.

7.1.3 Construction details

Construction details for columns in buildings are:

(1) beam-to-column connections;
(2) column cap connections;
(3) column splices;
(4) column bases.

(1) Beam-to-column cap connections

Typical beam-to-column connections and column cap connections are shown in Figures 7.3(a) and (b), respectively.

(2) Column splices

Splices in compression members are discussed in Section 6.1.8.2 of BS 5950: Part 1. The code states that where the members are not prepared for full contact in bearing, the splice should be designed to transmit all the moments and forces to which the member is subjected. Where the members are prepared for full contact, the splice should provide continuity of stiffness about both axes and resist any tension caused by bending.

(a)

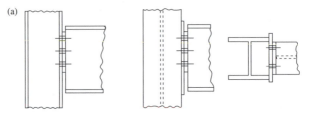

Flexible beam to column connections

(b)

Column cap connections

(c)

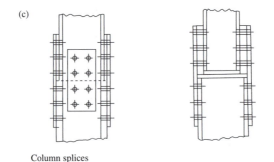

Column splices

Figure 7.3 Column construction details

In multi-storey buildings, splices are usually located just above floor level. If butted directly together, the ends are usually machined for bearing. Fully bolted splices and combined bolted and welded splices are used. If the axial load is high and the moment does not cause tension the splice holds the columns' lengths in position. Where high moments have to be resisted, high strength or friction-grip bolts or a full-strength welded splice may be required. Some typical column splices are shown in Figure 7.3(c).

(3) Column bases

Column bases are discussed in Section 7.10.

7.2 Loads on compression members

Axial loading on columns in buildings is due to loads from roofs, floors and walls transmitted to the column through beams and to self weight (see Figure 7.4(a)). Floor beam reactions are eccentric to the column axis, as shown,

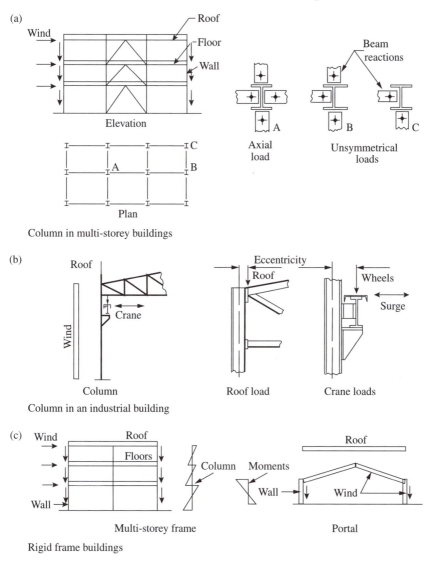

Figure 7.4 Loads and moments on compression members

and if the beam arrangement or loading is asymmetrical, moments are transmitted to the column. Wind loads on multi-storey buildings designed to the simple design method are usually taken to be applied at floor levels and to be resisted by the bracing, and so do not cause moments.

In industrial buildings, loads from cranes and wind cause moments in columns, as shown in Figure 7.4(b). In this case, the wind is applied as a distributed load to the column through the sheeting rails.

In rigid frame construction moments are transmitted through the joints from beams to column, as shown in Figure 7.4(c). Rigid frame design is outside the scope of this book.

7.3 Classification of cross-sections

The same classification that was set out for beams in Section 5.3 is used for compression members. That is, to prevent local buckling, limiting proportions for flanges and webs in axial compression are given in Table 11, BS 5950: Part 1. The proportions for rolled and welded column sections are shown in Figure 7.5.

7.4 Axially loaded compression members

7.4.1 General behaviour

Compression members may be classified by length. A short column, post or pedestal fails by crushing or squashing, as shown in Figure 7.6(a). The squash load P_y in terms of the design strength is:

$$P_y = p_y A$$

where A is the area of cross-section.

A long or slender column fails by buckling, as shown in Figure 7.6(b). The failure load is less than the squash load and depends on the degree of slenderness. Most practical columns fail by buckling. For example, a universal column under axial load fails in flexural buckling about the weaker y–y axis (see Figure 7.6(c)).

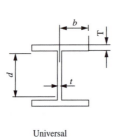

Universal
column

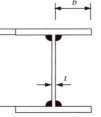

H-section
column

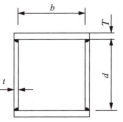

Box column

Limiting proportions

Element	Section Type	Class 1- Plastic Section	Class 2- Compact Section	Class 3- Semi Compact Section
Outstand element of compression flange	Rolled $b/T \leq$	9.0ε	10.0ε	15.0ε
	Welded $b/T \leq$	8.0ε	8.0ε	8.0ε
Internal element of compression flange	Welded $b/T \leq$	8.0ε	8.0ε	8.0ε
Web subject to compression throughout	Rolled $d/t \leq$	–	–	$120\varepsilon/(1+2r_2)$
	Welded $d/t \leq$			but $\varepsilon \leq 40\varepsilon$

All elements in compression due to axial load: $\varepsilon = (275/p_y)^{0.5}$; $r_2 = F_c/(A_g p_{yw})$

Figure 7.5 Limiting proportions for rolled and welded column sections

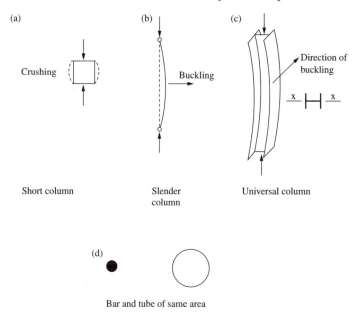

Figure 7.6 Behaviour of members in axial compression

The strength of a column depends on its resistance to buckling. Thus the column of tubular section shown in Figure 7.6(d) will carry a much higher load than the bar of the same cross-sectional area.

This is easily demonstrated with a sheet of A4 paper. Open or flat, the paper cannot be stand on edge to carry its own weight; but rolled into a tube, it will carry a considerable load. The tubular section is the optimum column section having equal resistance to buckling in all directions.

7.4.2 Basic strut theory

(1) Euler load

Consider a pin-ended straight column. The critical value of axial load P is found by equating disturbing and restoring moments when the strut has been given a small deflection y, as shown in Figure 7.7(a). The equilibrium equation is:

$$EI_y \frac{d^2 y}{dx^2} = -Py$$

This is solved to give the Euler or lowest critical load:

$$P_E = \pi^2 EI_y/L^2$$

In terms of stress, the equation is:

$$P_E = \frac{\pi^2 E}{(L/r_y)^2} = \frac{\pi^2 E}{\lambda^2}$$

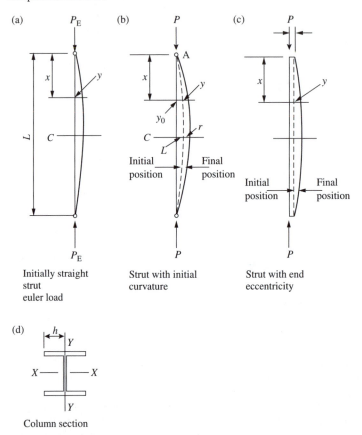

(a) P_E (b) P (c) P

Initially straight   Strut with initial   Strut with end
strut   curvature   eccentricity
euler load

(d)

Column section

Figure 7.7 Load cases for struts

where I_y is the moment of inertia about the minor axis y–y, L the length of the strut, P_P the axial load, r_y the radius of gyration for the minor axis y–y $= (I_y/A)^{0.5}$, $p_E = P_E/A =$ Euler critical stress and $\lambda =$ slenderness ratio $= L/r_y$.

The slenderness λ is the only variable affecting the critical stress. At the critical load, the strut is in neutral equilibrium. The central deflection is not defined and may be of unlimited extent. The curve of Euler stress against slenderness for a universal column section is shown in Figure 7.9.

(2) Strut with initial curvature

In practice, columns are generally not straight, and the effect of out of straightness on strength is studied in this section. Consider a strut with an initial curvature bent in a half sine wave, as shown in Figure 7.7(b). If the initial deflection at x from A is y_0 and the strut deflects y further under load P, the equilibrium equation is:

$$E I_y \frac{\mathrm{d}^2 y}{\mathrm{d}x^2} = P(y + y_0)$$

where deflection $y = \sin(\pi x / L)$. If δ_0 is the initial deflection at the centre and δ the additional deflection caused by P, then it can be shown by solving the equilibrium equation that:

$$\delta = \frac{\delta_0}{(P_E/P) - 1}$$

The maximum stress at the centre of the strut is given by:

$$P_{max} = \frac{P}{A} + \frac{P(\delta_0 + \delta)h}{I_Y}$$

where h is shown in Figure 7.7(d).
    In the above equation,

$$p_{max} = p_y = \text{design strength,}$$
$$p_c = P/A = \text{average stress,}$$
$$p_E = P_E/A = \text{Euler stress,}$$
$$I_y = Ar_y^2 = \text{moment of inertia about the } y\text{–}y \text{ axis,}$$
$$A = \text{area of cross-section,}$$
$$r_y = \text{radius of gyration for the } y\text{–}y \text{ axis,}$$
$$h = \text{half the flange breath.}$$

The equation for maximum stress can be written:

$$p_y = p_c + p_c \left\{ 1 + \frac{1}{(p_E/p_c) - 1} \right\} \frac{\delta_0 h}{r_y^2}$$

Put

$$\eta = \delta_0 h / r_y^2$$

and rearrange to give:

$$(p_E - p_c)(p_y - p_c) = \eta p_E p_c$$

The value of p_c the limiting strength at which the maximum stress equals the design strength, can be found by solving this equation and η is the Perry factor. This is to redefined in terms of slenderness. (See Section 7.4.3 (2) below. The design strength curve is also discussed in that section.)

(3) Eccentrically loaded strut

Most struts are eccentrically loaded, and the effect of this on strut strength is examined here. A strut with end eccentricities e is shown in Figure 7.7(c). If y is deflection from the initially straight strut the equilibrium equation is:

$$E I_y \frac{d^2 y}{dx^2} = -p(e + y)$$

This can be solved to give the secant formula for limiting stress.

Theoretical studies and tests show that the behaviour of a strut with end eccentricity is similar to that of one with initial curvature. Thus the two cases can be combined with the Perry factor, taking account of both imperfections.

7.4.3 Practical strut behaviour and design strength

(1) Residual stresses

As noted above, in general, practical struts are not straight and the load is not applied concentrically. In addition, rolled and welded strut sections have residual stresses which are locked in when the section cools.

A typical pattern of residual stress for a hot-rolled H-section is shown in Figure 7.8. If the section is subjected to a uniform load, the presence of these stresses causes yielding to occur first at the ends of the flanges. This reduces the flexural rigidity of the section, which is now based on the elastic core, as shown in Figure 7.8(b). The effect on buckling about the y–y axis is more severe than for the x–x axis. Theoretical studies and tests show that the effect of residual stresses can be taken into account by adjusting the Perry factor η.

(2) Column tests and design strengths

An extensive column-testing programme has been carried out, and this has shown that different design curves are required for:

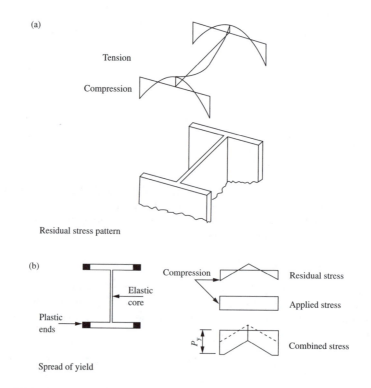

Figure 7.8 Residual stresses

(1) different column sections;
(2) the same section buckling about different axes;
(3) sections with different thicknesses of metal.

For example, H-sections have high residual compressive stresses at the ends of the flanges, and these affect the column strength if buckling takes place about the minor axis.

The total effect of the imperfections discussed above (initial curvature, end eccentricity and residual stresses on strength) are combined into the Perry constant η. This is adjusted to make the equation for limiting stress p_c a lower bound to the test results.

The constant η is defined by:

$$\eta = 0.001\, a(\lambda - \lambda_0)$$
$$\lambda = 0.2\, (\pi^2 E/p_y)^{0.5}$$

The value λ_0 gives the limit to the plateau over which the design strength p_y controls the strut load.

The Robertson constant a is assigned different values to give the different design curves. For H-sections buckling about the minor axis, a has the value 5.5 to give design curve (c) (Table 24(c)).

A strut table selection is given in Table 23 in BS 4950: Part 1. For example, for rolled and welded H sections with metal thicknesses up to 40 mm, the following design curves are used:

(1) buckling about the major axis $x–x$ curve (b) (Table 24(b));
(2) buckling about the minor axis $y–y$ curve (c) (Table 24(c)).

The compressive strength is given by the smaller root of the equation that was derived above for a strut with initial curvature. This is:

$$(p_E - p_c)(p_y - p_c) = \eta p_E p_c$$
$$p_c = \frac{p_E p_y}{(\phi + \phi^2 - p_E p_y)^{0.5}}$$
$$\phi = [p_y + (\eta + 1)p_E]/2$$

The curves for Euler stress p_E and limiting stress or compressive strength p_c for a rolled H-section column buckling about the minor axis are shown in

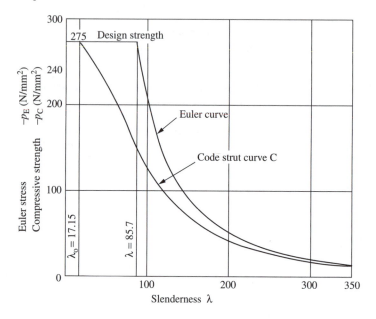

Figure 7.9 Strut strength curves

Figure 7.9. It can be noted that short struts fail at the design strength while slender ones approach the Euler critical stress. For intermediate struts, the compressive strength is a lower bound to the test results, as noted above. Compressive strengths for struts for curves a, b, c and d are given in Tables 24(a)–(d) in BS 5950: Part 1.

7.4.4 Effective lengths

(1) Theoretical considerations

The actual length of a compression member on any plane is the distance between effective positional or directional restraints in that plane. A positional restraint should be connected to a bracing system which should be capable of resisting 1% of the axial force in the restrained member. See Clause 4.7.1 of BS 5950.

The actual column is replaced by an equivalent pin-ended column of the same strength that has an effective length:

$$L_E = KL$$

where L is the actual length, and K the effective length ratio and K is to be determined from the end conditions.

An alternative method is to determine the distance between points of contraflexure in the deflected strut. These points may lie within the strut length or they may be imaginary points on the extended elastic curve. The distance so defined is the effective length.

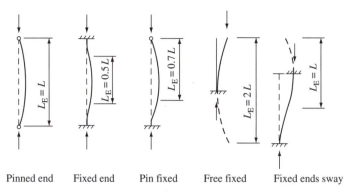

| Pinned end | Fixed end | Pin fixed | Free fixed | Fixed ends sway |

Figure 7.10 Figure effective lengths

The theoretical effective lengths for standard cases are shown in Figure 7.10. Note that for the cantilever and sway case the point of contraflexure is outside the strut length.

(2) Code definitions and rules

The effective length is defined in Section 1.2.14 of BS 5950: Part 1 as the length between points of effective restraint of a member multiplied by a factor to take account of the end conditions and loading.

Effective lengths for compression members are set out in Section 4.7.2 of the code. This states that for members other than angles, channels and T-sections, the effective length should be determined from the actual length and conditions of restraint in the relevant plane. The code specifies:

(1) That restraining members which carry more than 90 per cent of their moment capacity after reduction for axial load shall be taken as incapable of providing directional restraint.
(2) Table 22 is used for standard conditions of restraint.
(3) Appendix D1 is used for stanchions in single-storey buildings of simple construction (see Section 7.6).
(4) Appendix E is used for members forming part of a frame with rigid joints.

The normal effective lengths L_E are given in Table 22 of the code. Some values from this table for various end conditions where L is the actual length are:

(1) Effectively held in position at both ends
  (a) Restrained in direction at both ends, $L_E = 0.7L$
  (b) Partially restrained in direction.
    at both ends, $L_E = 0.85L$
  (c) Not restrained in direction at either end, $L_E = L$
(2) One end effectively held in position and restrained in direction. Other end not held in position
  (a) Partially restrained in direction, $L_E = 1.5L$
  (b) Not restrained in direction, $L_E = 2.0L$

The reader should consult the table in the code for other cases.

Note the case for the fixed end strut, where the effective length is given as $0.7\,L$, is to allow for practical ends where true fixity is rarely achieved. The theoretical value shown in Figure 7.10 is $0.5\,L$.

7.4.5 Slenderness

The slenderness λ is defined in Section 4.7.3 of the code as:

$$\lambda = \frac{\text{Effective length}}{\text{Radius of gyration about relevant axis}} = \frac{L_E}{r}$$

The code states that, for members resisting loads other than wind load, λ must not exceed 180. Wind load cases are dealt with in Chapter 8 of this book.

7.4.6 Compression resistance

The compression resistance of a strut is defined in Section 4.7.4 of BS 5950: Part 1 as:

(1) Plastic, compact or semi-compact sections: $P_c = A_g p_c$
(2) Slender sections: $P_c = A_{\text{eff}} p_{cs}$

where A_g is the gross sectional area defined in Section 3.4.1 of the code, A_{eff} the effective sectional area defined in Section 3.6.2 of the code, p_c the compressive strength from Section 4.7.5 and Tables 27(a)–(d) of the code and p_{cs} the value p_c from clause 4.75 for a reduced slenderness of $\lambda(A_{\text{eff}}/A_g)^{0.5}$ in which λ is based on the radius of gyration r of the gross cross-sections.

7.4.7 Column design

Column design is indirect, and the process is as follows (the tables referred to are in the code):

(1) The steel grade and section is selected.
(2) The design strength p_y, is taken from Table 9.
(3) The effective length L_E is estimated using Table 22 for the appropriate end conditions.
(4) The slenderness λ is calculated for the relevant axis.
(5) The strut curve is selected from Table 23.
(6) The compressive strength is read from the appropriate part of Tables 24(a)–(d).
(7) The compression resistance P_c is calculated (see Section 7.4.6 above).

For a safe design, P_c should just exceed the applied load, and successive trials are needed to obtain an economical design. Load tables can be formed to give the compression resistance for various sections for different values of effective length. Table 7.1 gives compression resistances for some universal column sections. Column sizes may be selected from tables in the *Guide* to BS 5950: Part 1, Volume 1, Section Properties, Member Capacities, Steel Construction Institute.

Table 7.1 Compression resistance of S275 steel U.C. sections

Serial size (mm)	Mass per metre (kg)	Compression resistances (kn) for effective lengths (m)								
		2	2.5	3	3.5	4	5	6	8	10
254 × 254	167	5230	4990	4730	4460	4180	3590	3010	2080	3580
universal	132	4160	3960	3750	3530	3300	2820	2360	1620	1140
column	107	3360	3200	3020	2840	2650	2260	1880	1280	909
	89	2790	2650	2510	2360	2200	1860	1550	1060	742
	73	2350	2230	2110	1970	1830	1550	1270	860	602
203 × 203	86	2570	2400	2220	2030	1830	1460	1150	740	–
universal	71	2130	1980	1830	1670	1510	1200	943	605	–
column	60	1820	1700	1560	1410	1270	995	778	494	–
	52	1600	1480	1360	1230	1100	865	676	1429	–
	46	1410	1310	1200	1080	968	757	590	374	–
152 × 152	37	1030	910	787	671	568	411	306	–	–
universal	30	825	727	627	533	450	325	241	–	–
column	23	632	552	472	397	334	239	177	–	–

7.4.8 Example: universal column

A part plan of an office floor and the elevation of internal column stack A are shown in Figures 7.11(a) and (b). The roof and floor loads are as follows:

Roof:
Dead load (total) $= 5 \, \text{kN/m}^2$;
Imposed load $= 1.5 \, \text{kN/m}^2$.

Floors:
Dead load (total) $= 7 \, \text{kN/m}^2$;
Imposed load $= 3 \, \text{kN/m}^2$

Design column *A* for axial load only. The self-weight of the column, including fire protection, may be taken as 1 kN/m. The roof and floor steel have the same layout. Use Grade S275 steel.

When calculating the loads on the column lengths, the imposed loads may be reduced in accordance with Table 2 of BS 6399: Part 1. This is permitted because it is unlikely that all floors will be fully loaded simultaneously. Values from the table are:

Number of floor carried by member	Reduction in imposed load (%)
1	0
2	10
3	20

The roof is regarded as a floor for reckoning purposes.

The slabs for the floor and roof are precast one-way spanning slabs. The dead and imposed loads are calculated separately.

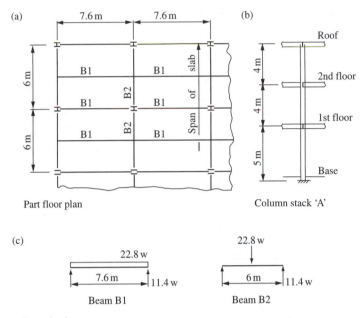

Beam loads

Figure 7.11 Column design example

(1) Loading

Four floor beams are supported at column A. These are designated as B1 and B2 in Figure 7.11 (a). The reactions from these beams in terms of a uniformly distributed load are shown in Figure 7.11(c):

$$\text{Load on beam B1} = 7.6 \times 3 \times 10 = 22.8 \, w \, (\text{kN})$$

where w is the uniformly distributed load. The dead and imposed loads must be calculated separately in order to introduce the different load factors. The self weight of beam B2 is included in the reaction from beam B1.

The design loading on the column can be set out as shown in Figure 7.12. The design loads are required just above the first floor, the second floor and the base.

(2) Column design

(1) Top length = Roof to second floor
    Design load = 434.2 kN
    Try 152 × 152 UC 30
    $A = 38.2 \, \text{cm}^2$; $r_y = 3.82 \, \text{cm}$
    Design strength $p_y = 275 \, \text{N/mm}^2$ (Table 9) where section thickness is less than 16 mm.

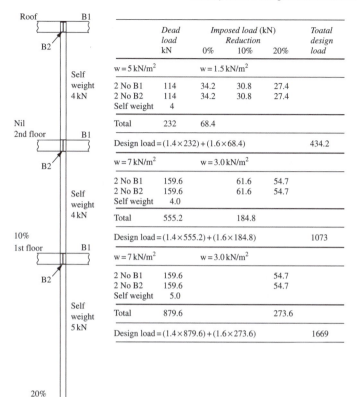

	Dead load kN	Imposed load (kN) Reduction			Toatal design load
		0%	10%	20%	
$w = 5\,kN/m^2$		$w = 1.5\,kN/m^2$			
2 No B1	114	34.2	30.8	27.4	
2 No B2	114	34.2	30.8	27.4	
Self weight	4				
Total	232	68.4			
Design load $= (1.4 \times 232) + (1.6 \times 68.4)$					434.2
$w = 7\,kN/m^2$		$w = 3.0\,kN/m^2$			
2 No B1	159.6		61.6	54.7	
2 No B2	159.6		61.6	54.7	
Self weight	4.0				
Total	555.2		184.8		
Design load $= (1.4 \times 555.2) + (1.6 \times 184.8)$					1073
$w = 7\,kN/m^2$		$w = 3.0\,kN/m^2$			
2 No B1	159.6			54.7	
2 No B2	159.6			54.7	
Self weight	5.0				
Total	879.6			273.6	
Design load $= (1.4 \times 879.6) + (1.6 \times 273.6)$					1669

Figure 7.12 Column design loads

If the beam connections are the shear type discussed in Section 5.8.3, where end rotation is permitted, the effective length, from Table 22:

$$L_E = 0.85 \times 4000 = 3400\,mm$$
Slenderness, $\lambda = 3400/38.2 = 89$

For a rolled H section thickness less than 40 mm buckling about the minor y–y axis, use Table 24, curve (c):

Compressive strength $p_c = 144\,N/mm^2$
Compressive resistance $p_c = 144 \times 38.2/10 = 550.1\,kN$

The column splice and floor beam connections at second-floor level are shown in Figure 7.13(a). The net section at the splice is shown in Figure 7.13(b) with 4 No. 22 mm diameter holes. The section is satisfactory.
(2) Intermediate length—first floor to second floor. Design load = 1073 kN. Try 203 × 203 UC 46.

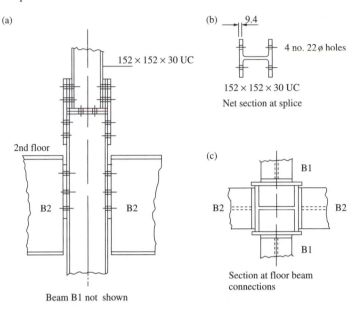

(a)

152 × 152 × 30 UC

2nd floor

B2   B2

Beam B1 not shown

Spice and floor beam connections

(b)   9.4

4 no. 22 ø holes

152 × 152 × 30 UC

Net section at splice

(c)

B1

B2   B2

B1

Section at floor beam
connections

Figure 7.13 Column connection details

$$A = 58.8 \, \text{cm}^2; \, r_y = 5.11 \, \text{cm}$$
$$p_y = 275 \, \text{N/mm}^2$$
$$\lambda = 3400/51.1 = 66.5$$
$$p_c = 188 \, \text{N/mm}^2$$
$$P_c = 188 \times 58.8/10 = 1105.4 \, \text{kN}$$

The section is satisfactory.
(3) Bottom length—base to first floor. Design load = 1669 kN.

Try 254 × 254 UC 73:
$A = 92.9 \, \text{cm}^2; \, r_y = 6.46 \, \text{cm}$.

The flange is 14.2 mm thick. The design strength from Table 9 in the code is

$$p_y = 275 \, \text{N/mm}^2$$

The beam connections do not restrain the column in direction at the first floor level. The base can be considered fixed. The effective length is taken as:

$$L_E = 0.85 \times 5000 = 4250 \, \text{mm}$$
$$\lambda = 4250/64.6 = 65.8$$

$$p_c = 189.4 \, \text{N/mm}^2 \, (\text{Table 24(c)})$$
$$P_c = 92.9 \times 189.4/10 = 1759.5 \, \text{kN}$$

The section selected is satisfactory. The same sections could have been selected from Table 7.1.

7.4.9 Built-up column: design

The two main types of columns built up from steel plates are the H and box sections shown in Figure 7.2. The classification for cross-sections is given in Figure 7.5.

For plastic, compact or semi-compact cross-sections, the local compression capacity is based on the gross section. The code states in Section 4.7.5 that the design strength p_y for sections fabricated by welding is to be the value from Table 9 reduced by 20 N/mm². This takes account of the severe residual stresses and possible distortion due to welding.

Slender cross-sections are dealt with in Section 3.6 of the code. The capacity of these sections is limited by local buckling and the design should be based on the effectively cross-sectional area. The compressive strength p_{cs} should be evaluated from Clause 4.75 with a reduced slenderness of $\lambda(A_{eff}/A_g)^{0.5}$.

7.4.10 Example: built-up column

Determine the compression resistance of the column section shown in Figure 7.14. The effective length of the column is 8 m and the steel is S275.

(1) Flanges

The design strength from Table 9 for plate 30mm thick $p_y = 265$ N/mm². Reducing by 20 N/mm² for a welded section gives:

$$p_y = 245 \text{ N/mm}^2$$
$$\varepsilon = (275/245)^{0.5} = 1.059$$

$$\text{Flange outstand } b = 442.5 = 14.75\,T$$
$$> 13\,\varepsilon T = 313.77\,T$$

Referring to Table 11, the flange is slender. The effective area per flange is:
$13\varepsilon T \times 2 \times T = 13 \times 1.059 \times 30 \times 2 \times 30 = 24\,780 \text{ mm}^2$.

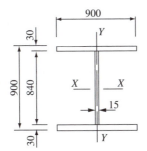

Figure 7.14 Built-up H column

(2) Web

This is an internal element in axial compression.

$$p_y = 275 - 20 = 255 \, \text{N/mm}^2$$
$$\varepsilon = (275/255)^{0.5} = 1.038$$

The effective area of the web is taken as $20\varepsilon t$ from each end, hence for the web effective area is $20 \times 1.038 \times 15 \times 2 \times 15 = 9342 \, \text{mm}^2$

(3) Properties of the gross section and effective section

Gross area $= (2 \times 30 \times 900) + (840 \times 15) = 66\,600 \, \text{mm}^2$

$I_y = (60 \times 900^3/12) + (\text{neglect web}) = 3.645 \times 10^9 \, \text{mm}^4$;

$r_y = [3.645 \times 10^9/6.66 \times 10^4]^{0.5} = 233.9 \, \text{mm}$;

$\lambda = 8000/233.9 = 34.2$.

Effective sectional area $= 2 \times 24\,780 + 9342 = 58\,902 \, \text{mm}^2$.

(4) Compressive resistance of the column

The compressive strength of the column p_c is obtained with p_y of 245 N/mm^2 and reduced slenderness of

$$\lambda (A_{\text{eff}}/A_{\text{g}})^{0.5} = 34.2 \times (58\,902/66\,600)^{0.5} = 32.2.$$

From Table 24c, p_c is 225 N/mm^2,

$$P_c = p_c A_{\text{eff}} = 225 \times 58\,902/1000 = 13\,253 \, \text{kN}.$$

This is compared with the strength of 12 300 kN calculated from the previous version of the code. Note that the new procedure is much easier.

7.4.11 Cased columns: design

(1) General requirements

Solid concrete casing acts as fire protection for steel columns and the casing assists in carrying the load and preventing the column from buckling about the weak axis. Regulations governing design are set out in Section 4.14 of BS 5950: Part 1.

The column must meet the following general requirements:

(1) The steel section is either a single-rolled or fabricated I- or H-section with equal flanges. Channels and compound sections can also be used. (Refer to the code for requirements.)
(2) The steel section is not to exceed $1000 \times 500 \, \text{mm}^2$. The dimension 1000 mm is in the direction of the web.
(3) Primary structural connections should be made to the steel section.

(4) The steel section is unpainted and free from dirt, grease, rust, scale, etc.
(5) The steel section is encased in concrete of at least Grade 25, to BS 8110.
(6) The cover on the steel is to be not less than 50 mm. The corners may be chamfered.
(7) The concrete extends the full length of the member and is thoroughly compacted.
(8) The casing is reinforced with steel fabric mesh #D98 per BS 4483 or alternatively with rebars not less than 5 mm diameter at a maximum spacing of 200 mm to form a cage of closed links and longitudinal bars. The reinforcement is to pass through the centre of the cover, as shown in Figure 7.15(a).
(9) The effective length is not to exceed $40b_c$, $100b_c^2/d_c$ or $250r$, whichever is the least, where b_c is the minimum width of solid casing, d_c the minimum depth of solid casing and r the minimum radius of gyration of the steel section.

(2) Compression resistance

The design basis set out in Section 4.14.2 of the code is as follows:

(1) The radius of gyration about the *y–y* axis shown in Figure 7.15, r_y should be taken as $0.2b_c$ but not more than $0.2(B + 150)$, where B is the overall width of the steel flange. The radius of gyration for the *x–x* axis r_x should be taken as that of the steel section.
(2) The compression resistance P_c is

$$P_c = \left(A_g + 0.45 \frac{f_{cu}}{p_y} A_c \right) p_c$$

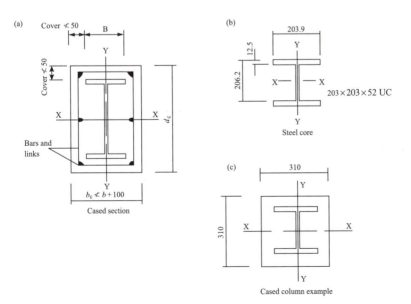

Figure 7.15 Cased column

but not more than the short strut capacity,

$$P_{cs} = \left(A_g + 0.25 \frac{f_{cu}}{p_y} A_c \right) p_y$$

where A_c is the gross sectional area of the concrete. Casing in excess of 75 mm from the steel section is neglected. Finish is neglected. A_g the gross area of the steel section, f_{cu} the characteristic strength of the concrete at 28 days. This is not to exceed 40 N/mm^2, p_c the compress strength of the steel section determined using r_x and r_y, in the determination of which $p_y < 355$ N/mm^2, and p_y the design strength of the steel.

7.4.12 Example: cased column

An internal column in a multi-storey building has an actual length of 4.2 m centre-to-centre of floor beams. The steel section is a 203 × 203 UC 52. Calculate the compression resistance of the column if it is cased in accordance with Section 4.14 of BS 5950: Part 1. The steel is Grade S275 and the concrete Grade 25. The steel core and cased section are shown in Figure 7.15(b). The casing has been made 310 mm^2.

The properties of the steel section are:

$$A = 66.4 \, \text{cm}^2, \quad r_x = 8.9 \, \text{cm}, \quad r_y = 5.16 \, \text{cm}$$

For the cased section:

$$r_y = 0.2 \times 310 = 62 \, \text{mm}$$
$$\geq 0.2(203.9 + 150) = 70.78 \, \text{mm}.$$

Because the column is cased throughout, the effective length is taken from Table 22 as 0.7 of the actual length: Effective length $L_E = 0.7 \times 4200 = 2940$ mm.

The effective length L_E is not to exceed:

$$40b_c = 40 \times 310 = 12\,400 \, \text{mm}$$
$$100b_c^2/d_c = 100 \times 310 = 3100 \, \text{mm}$$
$$250r = 250 \times 51.6 = 12\,900 \, \text{mm}$$
Slenderness, $\lambda = 2940/62 = 47.4$

The design strength from Table 9, $p_y = 275$ N/mm^2. For buckling about y–y, select curve (c) from Table 23. Compressive strength from Table 24(c):

$$p_c = 225.2 \, \text{N/mm}^2$$

The gross sectional area of the concrete:

$$A_c = 310 \times 310 = 96\,100 \, \text{mm}^2$$

Compressive resistance of the cased section:

$$P_c = \left(66.4 \times 10^2 + 0.45\frac{25 \times 96\,100}{275}\right)\frac{225.2}{10^3}$$
$$= 1495.3 + 885.3$$
$$= 2380.6\,\text{kN}$$

This is not to exceed the short strut capacity:

$$P_{cs} = \left(66.4 \times 10^2 + 0.25\frac{25 \times 96\,100}{275}\right)\frac{275}{10^3}$$
$$= 1826 + 600.6$$
$$= 2426.6\,\text{kN}$$

The compression resistance is 2380.6 kN.

7.5 Beam columns

7.5.1 General behaviour

(1) Behaviour classification

As already stated at the beginning of this chapter, most columns are subjected to bending moment in addition to axial load. These members, termed 'beam-columns', represent the general load case of an element in a structural frame. The beam and axially loaded column are limiting cases.

Consider a plastic or compact H-section column as shown in Figure 7.16(a). The behaviour depends on the column length, how the moments are applied and the lateral support, if any, provided. The behaviour can be classified into the following five cases:

Case 1: A short column subjected to axial load and uniaxial bending about either axis or biaxial bending. Failure generally occurs when the plastic capacity of the section is reached. Note limitations set in (2) below.

Case 2: A slender column subjected to axial load and uniaxial bending about the major axis x–x. If the column is supported laterally against buckling about the minor axis y–y out of the plane of bending, the column fails by buckling about the x–x axis. This is not a common case (see Figure 7.17(a)). At low axial loads or if the column is not very slender, a plastic hinge forms at the end or point of maximum moment

Case 3: A slender column subjected to axial load and uniaxial bending about the minor axis y–y. The column does not require lateral support and there is no buckling out-of-the-plane of bending. The column fails by buckling about the y–y axis. At very low axial loads, it will reach the bending capacity for y–y axis (see Figure 7.17(b)).

Case 4: A slender column subjected to axial load and uniaxial axial bending about the major axis x–x. This time the column has no lateral support. The column fails due to a combination of column buckling aboutthe y–y axis

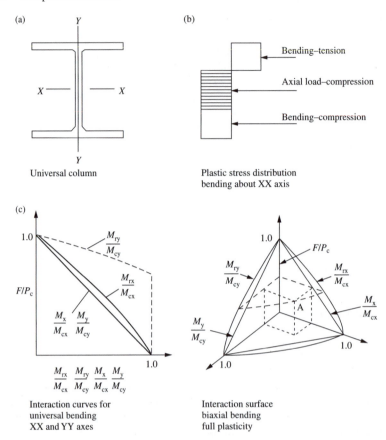

(a)

Universal column

(b)

Bending–tension

Axial load–compression

Bending–compression

Plastic stress distribution
bending about XX axis

(c)

$\dfrac{M_{ry}}{M_{cy}}$

$\dfrac{M_{rx}}{M_{cx}}$

$\dfrac{M_x}{M_{cx}} \quad \dfrac{M_y}{M_{cy}}$

F/P_c

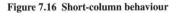

$\dfrac{M_{rx}}{M_{cx}} \ \dfrac{M_{ry}}{M_{cy}} \ \dfrac{M_x}{M_{cx}} \ \dfrac{M_y}{M_{cy}}$

Interaction curves for
universal bending
XX and YY axes

F/P_c

$\dfrac{M_{ry}}{M_{cy}}$

$\dfrac{M_y}{M_{cy}}$

A

$\dfrac{M_{rx}}{M_{cx}}$

$\dfrac{M_x}{M_{cx}}$

Interaction surface
biaxial bending
full plasticity

Figure 7.16 Short-column behaviour

and lateral torsional buckling where the column section twists as well as
deflecting in the x–x and y–y planes (see Figure 7.17(c)).

Case 5: A slender column subject to axial load and biaxial bending. The
column has no lateral support. The failure is the same as in Case 4 above
but minor axis buckling will usually have the greatest effect. This is the
general loading case (see Figure 7.17(d)).

Some of these cases are discussed in more detail below.

(2) Short-column failure

The behaviour of short columns subjected to axial load and moment is the
same as for tension members subjected to identical loads. This was discussed
in Section 7.3.3.

The plastic stress distribution for uniaxial bending is shown in
Figure 7.16(b). The moment capacity for plastic or compact sections in the

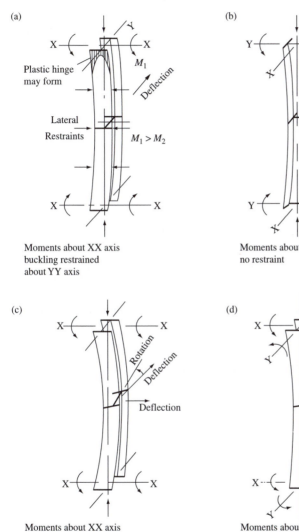

Figure 7.17 **Slender columns subjected to axial load and moment**

absence of axial load is given by:

$$M_c = Sp_y$$
$$\leq 1.2\, Zp_y \text{ (see Section 4.2.5 of BS 5950: Part 1)}$$

where S is the plastic modulus for the relevant axis and Z the elastic-modulus for the relevant axis.

The interaction curves for axial load and bending about the two principal axes separately are shown in Figure 7.18(a). Note the effect of the limitation of bending capacity for the y–y axis.

These curves are in terms of F/P_c against M_{rx}/M_{cx} and M_{ry}/M_{cy}, where F is the applied axial load, P_c the $p_y A$, the squash load, M_{rx} the reduced moment capacity about the x–x axis in the presence of axial load, M_{cx} the moment capacity about the x–x axis in the absence of axial load M_{ry} the reduced moment capacity about the y–y axis in the presence of axial load and M_{cy} the moment capacity about the y–y axis in the absence of axial load.

Values for M_{rx} and M_{ry} are calculated using equations for reduced plastic modulus given in the *Guide* to BS 5950: Part 1: Volume 1, Section Properties, Member Capacities, S.C.I.

Linear interaction expressions can be adopted. These are:

$$F/P_c + M_x/M_{cx} = 1$$

and

$$F/P_c + M_y/M_{cy} = 1$$

where M_x is the applied moment about the x–x axis and M_y the applied moment about the y–y axis. This simplification gives a conservative design.

Plastic and compact H-sections subjected to axial load and biaxial bending are found to give a convex failure surface, as shown in Figure 7.18(a). At any point A on the surface the combination of axial load and moments about the x–x and y–y axes M_x and M_y respectively, that the section can support can be read off.

A plane drawn through the terminal points of the surface gives a linear interaction expression.

$$\frac{F}{P_c} + \frac{M_x}{M_{cx}} + \frac{M_y}{M_{cy}} = 1$$

This results in a conservative design.

(3) Failure of slender columns

With slender columns, buckling effects must be taken into account. These are minor axis buckling from axial load and lateral torsional buckling from moments applied about the major axis. The effect of moment gradient must also be considered.

All the imperfections, initial curvature, eccentricity of application of load and residual stresses affect the behaviour. The end conditions have to be taken into account in estimating the effective length.

Theoretical solutions have been derived and compared with test results. Failure surfaces for H-section columns plotted from the more exact approach given in the code are shown in Figure 7.18(a) for various values of slenderness. Failure contours are shown in Figure 7.18(b). These represent lower bounds to exact behaviour.

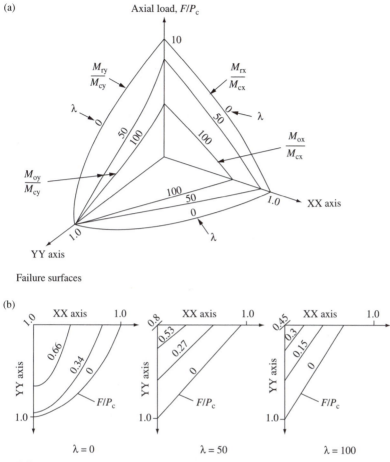

Figure 7.18 Failure surface for slender beam-column

The failure surfaces are presented in the following terms:

Slenderness $\lambda = 0$ $F_c/P_c; M_x/M_{cx}; M_y/M_{cy}$
$\lambda = 50, 100$ $F_c/P_c; M_{ax}/M_{cx}; M_{ay}/M_{cy}$

Some of the terms were defined in Section 7.5.1(2) above. New terms used are:

M_{ax} = maximum buckling moment about the x–x axis in the presence of axial load,

M_{ay} = maximum buckling moment about the y–y axis in the presence of axial load.

The following points are to be noted.

(1) M_{cy}, the moment capacity about the y–y axis, is not subjected to the restriction $1.2p_y Z_y$.

(2) At zero axial load, slenderness does not affect the bending strength of an H section about the y–y axis.
(3) At high values of slenderness the buckling resistance moment M_b about the x–x axis controls the moment capacity for bending about that axis.
(4) As the slenderness increases, the failure curves in the F/P_c, y–y-axis plane change from convex to concave, showing the increasing dominance of minor axis buckling.

For design purposes, the results are presented in the form of an interaction expression, and this is discussed in the next section.

7.5.2 Code design procedure

The code design procedure for compression members with moments is set out in Section 4.8.3 of BS 5950: Part 1. This requires the following two checks to be carried out:

(1) cross-section capacity check and
(2) member buckling check.

In each case, two procedures are given. These are a simplified approach and a more exact one. Only the simplified approach will be used in design examples in this book.

(1) Cross-section capacity check

The member should be checked at the point of greatest bending moment and axial load. This is usually at the end, but it could be within the column height if lateral loads are also applied. The capacity is controlled by yielding or local buckling. (Local buckling was considered in Section 7.3.)

Except for Class 4 members, with the simplified approach, the interaction relationship for Classes 1, 2 and 3 members given in Section 4.8.3.2 of the code is:

$$\frac{F_c}{A_g p_y} + \frac{M_x}{M_{cx}} + \frac{M_y}{M_{cy}} \leq 1$$

where F is the applied axial load, A_g the gross cross-sectional area, M_x the applied moment about the major axis x–x, M_{cx} the moment capacity about the major axis x–x in the absence of axial load, M_y the applied moment about the minor axis y–y, and M_{cy} the moment capacity about the minor axis y–y in the absence of axial load.

Alternatively, a more rigorous interaction relationship for plastic and compact sections given in the code can also be used. This is based on the convex failure surface discussed above and gives greater economy in design.

For Class 4 members, the interaction relationship is:

$$\frac{F_c}{A_{eff} p_y} + \frac{M_x}{M_{cx}} + \frac{M_y}{M_{cy}} \leq 1$$

where the additional term A_{eff} is the effective cross-sectional area defined by the code under Clause 3.6.

(2) Member buckling resistance

Under Clause 4.8.3.3.1 of the code, for simplified method, the buckling resistance of a member may be verified by checking the following relationships so that both are satisfied:

$$\frac{F_c}{A_g p_c} + \frac{m_x M_x}{M_b} + \frac{m_y M_y}{p_y Z_y} \leq 1$$

and

$$\frac{F_c}{A_g p_{cy}} + \frac{m_{LT} M_{LT}}{M_b} + \frac{m_y M_y}{p_y Z_y} \leq 1$$

where F_c = axial compressive load
   m = equivalent uniform moment factor (x or y axis) from Table 18 of the code,
   M_b = buckling resistance moment capacity about the major axis x–x,
   M_{LT} = the maximum major axis moment in the segment length L governing M_b,
   M_x = the maximum major axis moment in the segment length L_x govering p_{cx},
   M_y = the maximum minor axis moment in the segment length L_y governing p_{cy},
   p_c = the smaller of p_{cx} and p_{cy},
   p_{cy} = the compression resistance, considering buckling about the minor axis only,
   Z_y = elastic modulus of section for the minor axis y–y,
   Z_y = elastic modulus of section for the minor axis y–y.

The value for M_b is determined using the methods set out in Section 5.5 of this book (dealing with lateral torsional buckling of beams). A more exact approach is also given in the code. This uses the convex failure surfaces discussed above.

7.5.3 Example of beam column design

A braced column 4.5 m long is subjected to the factored end loads and moments about the x–x axis, as shown in Figure 7.19(a). The column is held in position but only partially restrained in direction at the ends. Check that a 203 × 203 UC 52 in Grade 43 steel is adequate.

(1) Column-section classification

Design strength from Table 9 $p_y = 275$ N/mm^2
Factor $\varepsilon = (275/p_y)^{0.5} = 1.0$
(see Figure 7.19(b))
Flange $b/T = 101.95/12.5 = 8.156 < 9.0$
Web $d/t = 160.8/8.0 = 20.1 < 40$

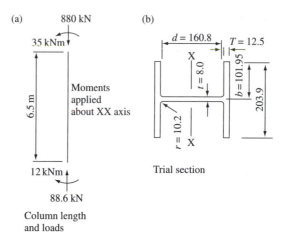

Figure 7.19 Beam column design example

Referring to Table 11, the flanges are plastic and the web semi-compact.

(2) Cross-section capacity check

Section properties for 203 × 203 UC 52 are:

$$A = 66.4\,\text{cm}^2; \quad Z_x = 510\,\text{cm}^3; \quad r_y = 516\,\text{cm}$$
$$x = 15.8; \quad u = 0.848; \quad S_x = 568\,\text{cm}^3$$

Moment capacity about the x–x axis:

$$M_{cx} = 275 \times 568/193 = 156.2\,\text{kN m}$$
$$< 1.2 \times 275 \times 510/10^3 = 168.4\,\text{kN m}$$

Interaction expression:

$$\frac{880 \times 10}{66.4 \times 275} + \frac{35}{156.2} = 0.48 + 0.22 = 0.7 < 1$$

The section is satisfactory with respect to local capacity.

$$= 0.48 + 0.22 = 0.7 < 1$$

(3) Member buckling check

The effective length from Table 22:

$$L_E = 0.85 \times 4500 = 3825$$
Slenderness $\lambda = 3825/51.6 = 74.1$

From Table 23, select Table 24(c) for buckling about the y–y axis. From Table 24(c), compressive strength $p_y = 172.8\,\text{N/mm}^2$.

Referring to Table 13, the support conditions for the beam column are that it is laterally restrained and restrained against torsion but partially free to rotate in plan:

Effective length $L_E = 0.85 \times 4500 = 3825\,\text{mm}$
Slenderness $\lambda = 74.1$

The ratio of end moments:

$$\beta = 12/35 = 0.342$$

From Table 18 the equivalent uniform moment factor $m_x = 0.697$.

$$\lambda_{LT} = uv\lambda$$

where $u = 0.848$ and denotes buckling parameter for H-section, $N = 0.5$ for uniform section with equal flanges, $x = 15.8$, the torsional index, $\lambda/x = 74.1/15.8 = 4.69$, $v = 0.832$, the slendemess factor from Table 19, $\lambda_{LT} = 0.848 \times 0.832 \times 74.1 = 52.2$

From Table 16, the bending strength:

$$p_b = 232.7\,\text{N/mm}^2$$

Buckling resistance moment:

$$M_b = 232.7 \times 568/10^3 = 132.1\,\text{kN m}$$

Interaction expression:

$$\frac{F_c}{A_g p_c} + \frac{m_x M_x}{M_b} + \frac{m_y M_y}{p_y Z_y} \leq 1$$

$$\frac{880 \times 10}{172.8 \times 66.4} + \frac{0.697 \times 35}{132.1} + 0 = 0.77 + 0.18 = 0.95 < 1.0$$

The section is also satisfactory with respect to overall buckling.

7.6 Eccentrically loaded columns in buildings

7.6.1 Eccentricities from connections

The eccentricities to be used in column design in simple construction for beam and truss reactions are given in Clause 4.7.6 of BS 5950: Part 1. These are as follows:

(1) For a beam supported on a cap plate, the load should be taken as acting at the face of the column or edge of the packing.
(2) For a roof truss on a cap plate, the eccentricity may be neglected provided that simple connections are used.

(3) In all other cases, the load should be taken as acting at a distance from the face of the column equal to 100 mm or at the centre of the stiff bearing, whichever gives the greater eccentricity.

The eccentricities for the various connections are shown in Figure 7.20.

7.6.2 Moments in columns of simple construction

The design of columns is set out in Section 4.7.7 of the code. The moments are calculated using eccentricities given in Section 7.6.1 above. For multi-storey columns effectively continuous at splices, the net moment applied at any one level may be divided between lengths above and below in proportion to the stiffness I/L of each length. When the ratio of stiffness does not exceed 1.5, the moments may be divided equally. These moments have no effect at levels above or below that at which they are applied.

The following interaction equation should be satisfied for the overall buckling check:

$$\frac{F_c}{A_g p_c} + \frac{M_x}{M_{bs}} + \frac{M_y}{p_y Z_y} \le 1$$

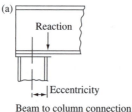

(a)

Reaction

Eccentricity

Beam to column connection

(b) Reaction

Truss to column connection

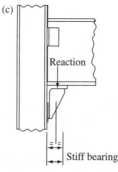

(c)

Reaction

Stiff bearing

Beam supported on bracket

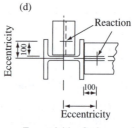

(d)

Reaction

Eccentricity

100

100

Eccentricity

Eccentricities for beam-column connections

Figure 7.20 Eccentricies for end reactions

where M_{bs} is the buckling resistance moment for a simple column calculated using an equivalent slenderness, $\lambda_{LT} = 0.5L/r_y$, I the moment of inertia of the column about the relevant axis, L the distance between levels at which both axes are restrained, r_y the radius of gyration about the minor axis and F_c the compressive force in the column.

Other terms are defined in Section 7.5.2.

7.6.3 Example: corner column in a building

The part plan of the floor and roof steel for an office building is shown in Figure 7.21(a) and an elevation of the corner column is shown in Figure 7.21(b). The roof and floor loading is as follows:

Roof:
Total dead load $= 5 \, \text{kN/m}^2$
Imposed load $= 1.5 \, \text{kN/m}^2$
Floors:
Total dead load $= 7 \, \text{kN/m}^2$
Imposed load $= 3 \, \text{kN/m}^2$

The self-weight of the column, including fire protection, is 1.5 kN/m. The external beams carry the following loads due to brick walls and concrete casing (they include self-weight):

Roof beams—parapet and casing $= 2 \, \text{kN/m}$
Floor beams—walls and casing $= 6 \, \text{kN/m}$

The reinforced concrete slabs for the roof and floors are one-way slabs spanning in the direction shown in the figure.

Design the corner column of the building using S275 steel. In accordance with Table 2 of BS 6399: Part 1, the imposed loads may be reduced as follows:

One floor carried by member—no reduction
Two floors carried by member—10% reduction
Three floors carried by member—20% reduction

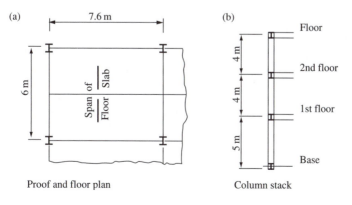

(a)   7.6 m   (b)

6 m

Span of Floor Slab

4 m   Floor
       2nd floor
4 m
       1st floor
5 m
       Base

Proof and floor plan   Column stack

Figure 7.21 Corner-column design example

The roof is counted as a floor. Note that the reduction is only taken into account in the axial load on the column. The full imposed load at that section is taken in calculating the moments due to eccentric beam reactions.

(1) Loading and reactions floor beams

Mark numbers for the floor beams are shown in Figure 7.22(a). The end reactions are calculated below:

Roof
B1   Dead load $= (5 \times 3.8 \times 1.5) + (2 \times 3.8) = 36.1\,\text{kN}$
      Imposed load $= 1.5 \times 3.8 \times 1.5 = 8.55\,\text{kN}$
B2   Dead load $= 5 \times 3.8 \times 3 = 57.0\,\text{kN}$
      Imposed load $= 1.5 \times 3.8 \times 3 = 17.1\,\text{kN}$
B3   Dead load $= (0.5 \times 57.0) + (2 \times 3) = 34.5\,\text{kN}$
      Imposed load $= 0.5 \times 17.1 = 8.55\,\text{kN}$

Floors
B1   Dead load $= (7 \times 3.8 \times 1.5) + (6 \times 3.8) = 62.7\,\text{kN}$
      Imposed load $= 3 \times 3.8 \times 1.5 = 17.1\,\text{kN}$
B2   Dead load $= 7 \times 3.8 \times 3 = 79.8\,\text{kN}$
      Imposed load $= 3 \times 3.8 \times 3 = 34.2\,\text{kN}$
B3   Dead load $= (0.5 \times 79.8) + (6 \times 3) = 57.9\,\text{kN}$
      Imposed load $= 0.5 \times 34.2 = 17.1\,\text{kN}$

The roof and floor beam reactions are shown in Figure 7.22(b).

(2) Loads and moments at roof and floor levels

The loading at the roof, second floor, first floor and base is calculated from values shown in Figure 7.22(b). The values for imposed load are calculated separately, so that reductions permitted can be made and the appropriate load factors for dead and imposed load introduced to give the design loads and moments.

The moments due to the eccentricities of the roof and floor beam reactions are based on the following assumed sizes for the column lengths:

    Roof to second floor:
        152 × 152 UC where the inertia I is proportional to 1.0;
    Second floor to first floor:
        203 × 203 UC where the inertia I is proportional to 2.5;
    First floor to base:
        203 × 203 UC where the inertia I is proportional to 2.5.

Further, it will be assumed initially that the moments at the floor levels can be divided between the upper and lower column lengths in proportion to the stiffnesses which are based on the inertia ratios given above. The actual values are not required.

The division of moments is made as follows:

(1) Joint at second floor level
      Upper column length—stiffness $I/L = 1/4 = 0.25$
      Lower column length—stiffness $I/L = 2.5/4 = 0.625$

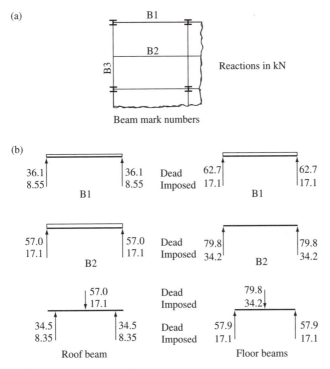

Figure 7.22 Floor–beam reactions

If M is the moment due to the eccentric floor beam reaction then the moment in the upper column length is:

$$M_u = [0.25/(0.25 + 0.625)]M = 0.286\,M$$

Moment in the lower column length is:

$$M_l = (1 - 0.286)M = 0.714\,M$$

(2) Joint at first level
It will be assumed that the same column section will be used for the two lower lengths. Hence the moments of inertia are the same and the stiffnesses are inversely proportional to the column lengths.
Upper column length—stiffness $= 1/4 = 0.25$
Lower column length—stiffness $= 1/5 = 0.20$
The stiffness of the upper column length does not exceed 1.5 times the stiffness of the lower length. Thus the moments may be divided equally between the upper and lower lengths.

The eccentricities of the beam reactions and the column loads and moments from dead and imposed loads are shown in Figure 7.23.

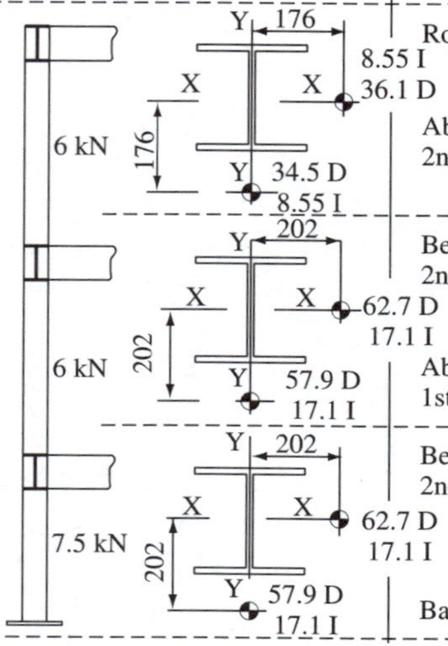

Column stack	Column sections	Position	Dead load	Imposed load	Reduced imp. load	Dead M_x	Imposee M_x	Dead M_y	Imposed M_y
Roof ×Nil I = 1.0		Roof	70.6	17.1	17.1	6.07	1.57	6.35	1.51
		Above 2nd floor	76.6	17.1	17.1	3.34	0.99	3.52	0.99
2nd floor ×10% I = 2.5		Below 2nd floor	197.2	31.3	46.2	8.35	2.47	9.01	2.47
		Above 1st Floor	203.2	31.3	46.2	5.84	1.79	6.33	1.73
1st floor ×20% I = 2.5		Below 2nd floor	323.8	85.5	68.4	5.84	1.79	6.33	1.73
		Base	331.3	68.4	—	—	—	—	—

Column sections:
Roof: Y 176, 8.55 I, 36.1 D, X, 34.5 D, 8.55 I, X 176, 6 kN
2nd floor: Y 202, 62.7 D, 17.1 I, X, 57.9 D, 17.1 I, X 202, 6 kN
1st floor: Y 202, 62.7 D, 17.1 I, X, 57.9 D, 17.1 I, X 202, 7.5 kN

× 10% permitted values for reduction in imposed loads
loads are in kN. moments in kNm

Figure 7.23 Loads and moments from actual and imposed loads

(3) Column design

Roof to second floor:
Referring to Figure 7.23, the design load and moments at roof level are:

$$\text{Axial load } F = (1.4 \times 70.6) + (1.6 \times 17.1) = 126.2\,\text{kN}$$
$$\text{Moment} \quad M_x = (1.4 \times 6.07) + (1.6 \times 1.51) = 10.92\,\text{kN m}$$
$$M_y = (1.4 \times 6.35) + (1.6 \times 1.51) = 11.32\,\text{kN m}$$

Try 152×152 UC 30, the properties of which are:

$$A = 38.2\,\text{cm}^2; \quad r_y = 3.82\,\text{cm}; \quad Z_x = 221.2\,\text{cm}^3;$$
$$Z_y = 73.06\,\text{cm}^3; \quad S_x = 247.1\,\text{cm}^3; \quad S_y = 111.2\,\text{cm}^3.$$

The roof beam connections and column section dimensions are shown in Figure 7.24(a).

The design strength from Table 9, $p_y = 275\,\text{N/mm}^2$, Flange, $b/T = 76.45/9.4 = 8.13 < 9.0$—plastic, Web, $d/T = 123.4/6.6 = 18.7 < 40$—semi-compact.

The limiting proportions are from Table 11 of the code.

Local capacity check:
Moment capacities for the *x–x* and *y–y* axes are:

$$M_{cx} = 247.1 \times 275/10^3 = 67.95\,\text{kN m},$$
$$< 1.2 \times 221.2 \times 275/10^3 = 73.0\,\text{kN m}.$$
$$M_{cy} = 111.2 \times 275/10^3 = 30.58\,\text{kN m},$$
$$< 1.2 \times 275 \times 73.06/10^3 = 24.10\,\text{kN m}.$$

Interaction expression:

$$\frac{126.2 \times 10}{38.2 \times 275} + \frac{10.92}{67.95} + \frac{11.31}{24.10} = 0.75 < 1.0$$

The section is satisfactory.
Overall buckling check:
The column is effectively held in position and partially restrained in direction at both ends. From Table 22, the effective length is:

$$L_E = 0.85 \times 4000 = 3400\,\text{mm}$$
$$\lambda = 3400/38.2 = 89$$

From Table 24(c)

$$p_c = 144\,\text{N/mm}^2$$

The axial load at the centre of the column is $= 126.2 + (3 \times 1.4) = 130.4\,\text{kN}$

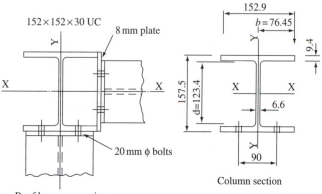

152×152×30 UC

8 mm plate

20 mm φ bolts

Roof beam connections

Column section

a) Column – roof to second floor

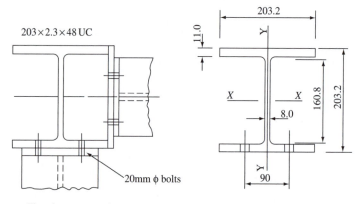

203×2.3×48 UC

20mm φ bolts

Floor beam connections

b) Column – second floor to base

Figure 7.24 Column connections and section dimensions

The buckling resistance moment M_b is calculated using Section 4.77 of the code:

$$\lambda_{LT} = 0.5 \times 4000/38.2 = 52.35$$
$$p_b = 232.2 \, \text{N/mm}^2 \, \text{(Table 16)}$$
$$M_b = 232.2 \times 247.1/101 = 57.4 \, \text{kN m}$$

Interaction expression:

$$\frac{130.4 \times 10}{38.2 \times 144} + \frac{10.92}{57.4} + \frac{11.32 \times 10^3}{275 \times 73.06} = 0.98 < 1.0.$$

The section is satisfactory.

Second floor to base:
The same column section will be used from the second floor to the base. The lower column length between first floor and base will be designed.

Referring to Figure 7.23, the design load and moments just below first floor level are:

$$F = (1.4 \times 323.8) + (1.6 \times 68.4) = 562.76 \text{ kN},$$
$$M_x = (1.4 \times 5.84) + (1.6 \times 1.73) = 10.94 \text{ kN m},$$
$$M_y = (1.4 \times 6.33) + (1.6 \times 1.73) = 11.63 \text{ kN m}.$$

Try 203×203 UC 46, the properties of which are:

$$A = 58.8 \text{ cm}^2; \quad r_y = 5.11 \text{ cm};$$
$$Z_y = 151.5 \text{ cm}^3; \quad S_x = 479.4 \text{ cm}^3.$$

Local capacity check:
The floor beam connections and column section dimensions are shown in Figure 7.24(b). The section is plastic and

$$p_y = 275 \text{ N/mm}^2$$

The moment capacities are:

$$M_{cy} = 275 \times 497.4/10^3 = 136.8 \text{ kN m},$$
$$M_{cy} = 1.2 \times 151.5 \times 275/10^3 = 50.0 \text{ kN m}.$$

Interaction expression:

$$\frac{562.76 \times 10}{58.8 \times 275} + \frac{10.94}{36.8} + \frac{11.63}{50.0} = 0.66 < 1.0.$$

Overall buckling check:

$$\lambda = 0.85 \times 5000/51 = 83.2$$
$$p_c = 155.2 \text{ N/mm}^2 \text{ (Table 24(c))}$$

Axial load at centre of column:

$$= 562.76 + (1.4 \times 3.75) = 568.01 \text{ kN},$$
$$\lambda_{LT} = 0.5 \times 5000/51.1 = 48.9.$$
$$p_b = 240.6 \text{ N/mm}^2 \text{—Table 11}.$$
$$M_b = 240.6 \times 497.4/10^3 = 119.6 \text{ kN m}.$$

Interaction expression:

$$\frac{568.01 \times 10}{58.8 \times 155.1} + \frac{10.94}{119.6} + \frac{11.63 \times 10^3}{275 \times 151.5} = 0.992 < 1.0.$$

The section is satisfactory.

7.7 Cased columns subjected to axial load and moment

7.7.1 Code design requirements

The design of cased members subjected to axial load and moment is set out in Section 4.14.4 of BS 5950: Part 1. The member must satisfy two conditions.

(1) Capacity check

$$\frac{F_c}{P_{cs}} + \frac{M_x}{M_{cx}} + \frac{M_y}{M_{cy}} \leq 1$$

where F_c is the compressive force due to axial load, P_{cs} the compressive resistance of a cased strut with zero slenderness(see Section 7.4.1 1), M_x the applied moment about the x–x axis and M_y the applied moment about the y–y axis, M_{cx} the moment capacity of the steel section about the x–x axis, and M_{cy} the moment capacity of the steel section about the y–y axis.

(2) Buckling resistance

$$\frac{F_c}{P_c} + \frac{m_x M_x}{M_b} + \frac{m_y M_y}{M_{cy}} \leq 1$$

where P_c is the compression resistance (see Section 7.4.11), m the equivalent uniform moment factor and M_b the buckling resistance moment calculated using the radius of gyration r_y for a cased section.

7.7.2 Example

A column of length 7 m is subjected to the factored loads and moments as shown in Figure 7.25. Design the column using S275 steel and Grade 30 concrete.
    Try 203 × 203 UC 60, the properties of which are:

$$A = 75.8\,\text{cm}^2, \quad S_x = 652\,\text{cm}^3,$$
$$r_x = 8.96\,\text{cm}, \quad r_y = 5.19\,\text{cm},$$
$$u = 0.847, \quad x = 14.1$$

The section is plastic and design strength $p_y = 275\,\text{N/mm}^2$. The cased section $320 \times 320\,\text{mm}^2$ is shown in Figure 7.25(b).

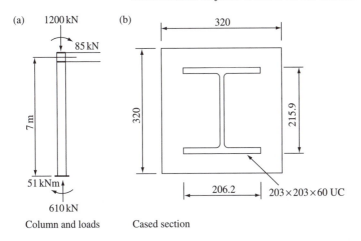

(a)   1200 kN   (b)

85 kN

7 m

51 kNm

610 kN

Column and loads   Cased section

Figure 7.25 Cased column: design example

(1) Capacity check

The terms for the interaction expression in Section 7.7.1(1) above are calculated:

$$P_{cs} = \left(75.8 + 0.25 \times \frac{30 \times 32^2}{275}\right) \frac{275}{10} = 2852 \text{ kN.}$$

$$M_{cx} = 652 \times 275/10^3 = 179.3 \text{ kN m.}$$

Interaction expression:

$$\frac{1200}{2852} + \frac{85}{179.3} = 0.42 + 0.474 = 0.894 < 1.0.$$

This is satisfactory.

(2) Buckling resistance

For the cased section $r_y = 0.2 \times 320 = 64$ mm. The strut is taken to be held in position and partially restrained in direction at the ends:

$$L_E = 7000 \times 0.85 = 5950 \text{ mm (Table 22).}$$

$$\lambda = 5950/64 = 93.0.$$

$$P_c = 137 \text{ N/mm}^2 \text{ (Table 24(c)).}$$

$$P_c = \left(75.8 + \frac{0.45 \times 30 \times 32^2}{275}\right) \frac{137}{10} = 1727 \text{ kN.}$$

From Table 18, for $\beta = -51/85 = -0.6$:

$$m_{LT} = 0.44,$$
$$\lambda = 93.0 \text{ (same as above)},$$
$$\lambda/x = 93/14.1 = 6.59,$$
$$v = 0.746 \text{ (Table 19)},$$
$$\lambda_{LT} = 0.847 \times 0.746 \times 93 = 58.7,$$
$$P_b = 216.3\,\text{N/mm}^2 \text{ (Table 16)},$$
$$M_b = 216.3 \times 652/10^3 = 141\,\text{kN m}.$$

Interaction expression:

$$\frac{1200}{1727} + \frac{0.44 \times 85}{141} = 0.694 + 0.265 = 0.959 < 1.0.$$

The section is satisfactory.

7.8 Side column for a single-storey industrial building

7.8.1 Arrangement and loading

The cross-section and side elevation of a single-storey industrial building are shown in Figures 7.26(a) and (b). The columns are assumed to be fixed at the base and pinned at the top, and act as partially propped cantilevers in resisting lateral loads. The top of the column is held in the longitudinal direction by the eaves member and bracing, as shown on the side elevation.

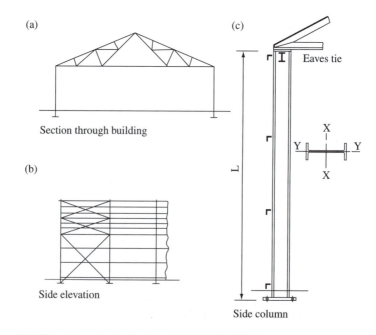

Figure 7.26 Side column in a single-storey industrial building

The loading on the column is due to:

(1) dead and imposed load from the roof and dead load from the walls and column; and
(2) wind loading on roof and walls.

The load on the roof consists of:

(1) dead load due to sheeting, insulation board, purlins and weight of truss and bracing. This is approximately 0.3–$0.5\,\text{kN/m}^2$ on the slope length of the roof; and
(2) Imposed load due to snow, erection and maintenance loads. This is given in BS 6399: Part 1 as $0.75\,\text{kN/m}^2$ on plan area.

The loading on the walls is due to sheeting, insulation board, sheeting rails and the weight of the column and bracing. The weight is approximately the same as for the roof.

The wind load depends on the location and dimensions of the building. The method of calculating the wind load is taken from CP3: Chapter V: Part 2. This is shown in the following example.

The breakdown and diagrams for the calculation of the loading and moments on the column are shown in Figure 7.27, and the following comments are made on these figures.

(1) The dead and imposed loads give an axial reaction R at the base of the column (see Figure 7.27(a)).
(2) The wind on the roof and walls is shown in Figure 7.27(b). There may be a pressure or suction on the windward slope, depending on the angle of the slope. The reactions from wind on the roof only are shown in Figure 7.27(c). The uplift results in vertical reactions R_1, and R_2. The net horizontal reaction is assumed to be divided equally between the two columns. This is $0.5(H_2 - H_1)$, where H_2 and H_1 are the horizontal components of the wind loads on the roof slopes.
(3) The wind on the walls causes the frame to deflect, as shown in Figure 7.27(d). The top of each column moves by the same amount S. The wind p, and P2 on each wall, taken as uniformly distributed, will have different values, and this results in a force P in the bottom chord of the truss, as shown in Figure 7.27(e). The value of P may be found by equating deflections at the top of each column. For the case where p_1, is greater than P2 there is a compression P in the bottom chord:

$$\frac{p_1 L^4}{8EI} - \frac{PL^3}{3EI} = \frac{p_2 L^4}{8EI} + \frac{PL^3}{3EI}$$

This gives

$$P = 3L(p_1 - p_2)/16.$$

where I denotes the moment of inertia of the column about the x–x axis (same for each column), E the Young's modulus and L the column height.

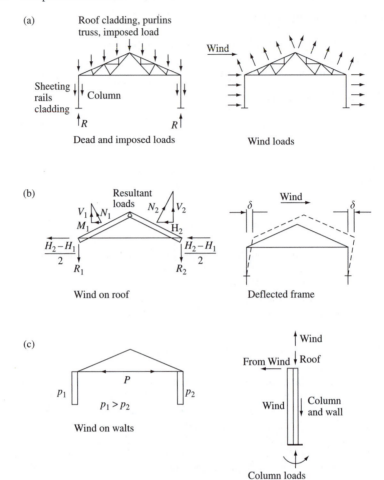

Figure 7.27 Loads on side column of an industrial building

(4) The resultant loading on the column is shown in Figure 7.27(f), where the horizontal point load at the top is:

$$H = P + (H_2 - H_1)/2$$

The column moments are due entirely to wind load.

7.8.2 Column design procedure

(1) Section classification

Universal beams are often used for these columns where the axial load is small, but the moment due to wind is large. Referring to Figure 7.28(a), the classification is checked as follows:

(1) Flanges are checked using Table 11 of the code where limits for b/T are given, where b is the flange outstand as shown in the figure and T is the flange thickness.
(2) Webs are in combined axial and flexural compression. The classification can be checked using Table 11 and Section 3.5.4 of the code. For example, from Table 11 for webs generally a plastic section has the limit:

$$\frac{d}{t} \leq \frac{80\varepsilon}{1 + r_1} \text{ but } \geq 40\varepsilon$$

where d is the clear depth of web, t the thickness of web and r_1 the stress ratio as defined in Clause 3.55 of the code.

(2) Effective length for axial compression

Effective lengths for cantilever columns connected by roof trusses are given in Appendix D of BS 5950: Part 1. The tops must be held in position longitudinally by eaves members connected to a braced bay.

Two cases are shown in Figure 7.28(b):

(1) Column with no restraints:
   x–x axis $L_E = 1.5L$
   y–y axis $L_E = 0.85L$.
   If the base is not effectively fixed about the y–y axis: $L_E = 1.0L$

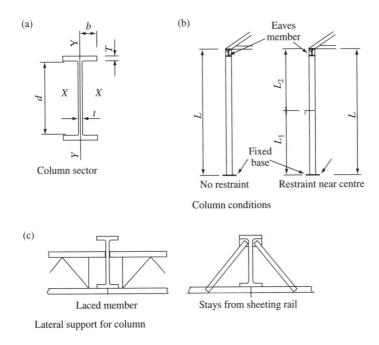

Figure 7.28 Side column design features

(2) Column with restraints:
The restraint provides lateral support against buckling about the weak axis:
x–x axis $L_E = 1.5L$
y–y axis $L_E = 0.85L_1$ or L_2, whichever is the greater.
The restraint is often provided by a laced member or stays from a sheeting rail, as shown in Figure 8.28(c).

(3) Effective length for calculating the buckling resistance moment

The effective length of compression flange is estimated using Sections 4.3.5, 4.3.6 and Tables 13 and 14 of BS 5950: Part 1, and the effective lengths for the two cases shown in Figure 7.28(b) are:

(1) Column with no restraints (Table 14):
The column is fixed at the base and restrained laterally and torsionally at the top. For normal loading $L_E = 0.5L$. Note that the code specifies in this case that the uniform moment factor m is taken as 1.0.
(2) Column with restraints:
This is to be treated as a beam and the effective length taken from Table 13:

$L_E = 0.85L_1$ or $1.0L_2$ in the case shown.

(4) Column design

The column moment is due to wind and controls the design. The load combination is then dead plus imposed plus wind load. The load factor from Table 2 of the code is $\gamma_f = 1.2$.
The following two checks are required in design:

(1) Local capacity check at base; and
(2) Overall buckling check.

The design procedure is shown in the example that follows.

(5) Deflection at the column cap

The deflection at the column cap must not exceed the limit given in Table 8 of the code for a single-storey building. The limit is height/300.

7.8.3 Example: design of a side column

A section through a single-storey building is shown in Figure 7.29. The frames are at 5 m centres and the length of the building is 30 m. The columns are pinned at the top and fixed at the base. The loading is as follows:

Roof
Dead load–measured on slope
Sheeting, insulation board, purlins and truss = $0.45\,\text{kN/m}^2$
Imposed load–measured on plan = $0.75\,\text{kN/m}^2$
Walls
Sheeting, insulation board, sheeting rails = $0.35\,\text{kN/m}^2$
Column
Estimate = 3.0 kN

Wind load: The new code of practice for wind loading is BS 6399, Part 2.

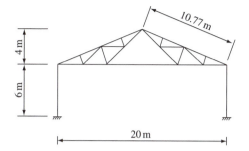

Figure 7.29 Section through building

Determine the loads and moments on the side column and design the member using S275 steel. Note that the column is taken as not being supported laterally between the top and base.

(1) Column loads and moments

Dead and imposed load

> *Roof*
> Dead load $= 10 \times 5 \times 0.45 \times 10.77/10 = 24.23$ kN
> Imposed load $= 10 \times 5 \times 0.75 = 37.5$ kN
> *Walls* $= 6 \times 5 \times 0.35 = 10.5$ kN
> Column $= 3.0$ kN
> Total load at base $= 75.23$ kN

Wind load

Location: north-east England. The site wind speed V_s is 26 m/s and the wind load factor S_b is 1.73 taken from Table 4, the effective wind speed, $V_e = V_s \times S_b = 45$ m/s.

The wind pressure coefficients and wind loads for the building are shown in Figure 7.30(b). The wind load normal to the walls and roof slope is given by:

$W = 5\,qL\,(C_{pe} - C_{pi})$,
$L =$ height of wall or length of roof slope,
$q =$ dynamic pressure for walls or roof slopes.

The resultant normal loads on the roof and the horizontal and vertical resolved parts are shown in Figure 7.30(c). The horizontal reaction is divided equally between each support and the vertical reactions are found by taking moments about supports. The reactions at the top of the columns for the two wind-load cases are shown in the figure.

The wind loading on the walls requires the analysis set out above in Section 7.8.1, where the column tops deflect by an equal amount and a force P is transmitted through the bottom chord of the truss. For the internal pressure case, see Figure 7.30(d):

$P = 3(8.31 - 6.64)/16 = 0.313$ kN

The loads and moments on the columns are summarized in Figure 7.31.

(a)

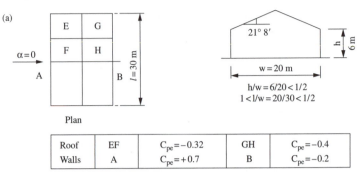

Plan

Roof	EF	$C_{pe}=-0.32$	GH	$C_{pe}=-0.4$
Walls	A	$C_{pe}=+0.7$	B	$C_{pe}=-0.2$

External pressure coefficients C_{pe} – Wind angle $\alpha=0$

(b)

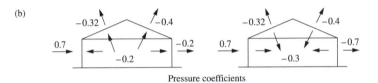

Pressure coefficients

Wind loads

Pressure coefficients and wind loads

(c)

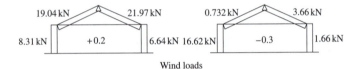

Roof loads and reactions

(d)

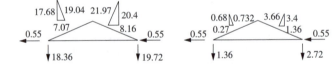

Wind on walls

Figure 7.30 Wind-pressure coefficients and loads

(2) Notional horizontal loads

To ensure stability, the structure is checked for a notional horizontal load in accordance with Clause 2.4.2.3 of BS 5950: Part 1. The notional force from the roof loads is taken as the greater of.

One per cent of the factored dead loads $= 0.01 \times 1.4 \times 24.23 = 0.34\,\text{kN}$ or 0.5 per cent of the factored dead load plus vertical imposed load $= 0.005\,1(1.4 \times 24.23) + (1.6 \times 37.5)1 = 0.5\,\text{kN}$.

This load is applied at the top of each column and is taken to act simultaneously with 1.4 times the dead and 1.3 times the imposed vertical loads.

Wind case	Internal pressure		Internal suction	
Column	Windward	Leeward	Windward	Leeward
Dead	↓24.23	↓24.23	↓24.23	↓24.23
Imposed	↓39.5	↓37.5	↓37.5	↓37.5
Wind	↓18.36	↑19.72	↑1.36	↑2.72
Wind wall column	.237 8.31 ↓13.5	.863 6.64 ↓13.5	2.87 16.6 ↓13.5	3.97 13.5↓ 1.66
Dead	↓37.73	↓37.73	↑37.73	↑37.73
Imposed	↓37.54	↓37.5	↑37.5	↑37.5
Wind	↓18.36	↑19.72	↓1.36	↓2.72
Wind moment	↲26.35	↲25.1	↲32.64	↲18.84

Loads are in kN, Moments are in kNm

Figure 7.31 Summary of loads and moments

The design load at the base is

$$P = (1.4 \times 37.73) + (1.3 \times 37.5) = 101.57\,\text{kN}.$$

The moment is:

$$M = 0.5 \times 6 = 3.0\,\text{kN m}.$$

The design conditions for this case are less severe than those for the combination dead + imposed + wind loads.

(3) Column design

The maximum design condition is for the wind-load case of internal suction. For the windward column, the load combination is dead plus imposed plus wind loads.

Local capacity check (see Section 7.52)

Design load $= 1.2(37.73 + 37.5 - 1.36) = 88.64\,\text{kN}$
Design moment $= 1.2 \times 32.64 = 39.17\,\text{kN m}$

Try 406×140 UB 39, the properties of which are:

$A = 49.4\,\text{cm}^2;\quad S_x = 721\,\text{cm}^3;\quad Z_x = 627\,\text{cm}^3,$
$r_x = 15.88\,\text{cm};\quad r_y = 2.89\,\text{cm};\quad x = 47.4,\quad I = 12\,452\,\text{cm}^4.$

Check the section classification using Table 7. The section dimensions are shown in Figure 7.32:

Design strength $p_y = 275\,\text{N/mm}^2$ (Table 9),
Factor $\varepsilon = 1.0$,
Flanges $b/T = 70.9/8.6 = 8.2 < 9.0$ (plastic),
Web. This is in combined axial and flexural compression.

Design axial load $= 88.64\,\text{kN}$,
Length of web supporting the load at the design strength:

$$= \frac{88.64 \times 10^3}{275 \times 6.3} = 51.1\,\text{mm},$$

For the web:

$$\frac{d}{t} = \frac{359.6}{6.3} = 57.1$$

Limiting value for plastic web:

$$\frac{d}{t} < \frac{80\varepsilon}{1 + r_1} \text{ but } \geq 40\varepsilon$$

where the web stress ratio r_1 is

$$r_1 = \frac{F_c}{dt\,P_{yw}} = \frac{88.64 \times 10^3}{359.6 \times 6.3 \times 275} = 0.1422,$$

$$\frac{d}{t} < \frac{80 \times 1.0}{1 + 0.1422} = 70 > 57.1 \text{ (plastic web)}$$

The moment capacity about the x–x axis:

$$M_{cx} = 275 \times 721/10^3 = 198.22\,\text{kN m}$$
$$< 1.2 \times 275 \times 627/10^3 = 206.9\,\text{kN m}$$

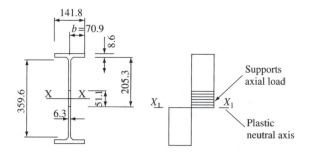

Figure 7.32 Column section

Interaction expression:

$$\frac{88.64 \times 10}{49.4 \times 275} + \frac{39.17}{198.22} = 0.27 < 1.0$$

The section is satisfactory.

(4) Overall buckling check

Compressive strength:

$$\lambda_x = 1.5 \times 6000/158.5 = 56.78,$$
$$\lambda_y = 0.85 \times 6000/28.9 = 176.47 < 180,$$
$$P_c = 52.1 \, \text{N mm}^2 \, \text{(Table 24(c))}.$$

Buckling resistance moment:

$$L_E = 0.5 \times 6000 = 3000 \, \text{(Table 14)},$$
$$\lambda = 3000/28.9 = 103.8.$$

Use the conservative approach in Section 4.3.7.7 of the code:

$$P_b = 146.8 \, \text{N/mm}^2 \, \text{from Table 20 for}$$
$$\lambda = 103.8 \quad \text{and} \quad x = 47.4,$$
$$M_b = 146.8 \times 721/103 = 105.8 \, \text{kN m}.$$

Interaction expression:

$$\frac{88.64 \times 10}{49.4 \times 52.1} - \frac{39.17}{105.8} = 0.35 + 0.37 = 0.71 < 1.0$$

The section is satisfactory. The slenderness exceeds 180 for the next lightest section.
Deflection at column cap:
For the internal suction case:

$$\delta = \frac{16.62 \times 10^3 \times 6000^3}{8 \times 205 \times 10^3 \times 12\,452 \times 10^4} - \frac{2.87 \times 10^3 \times 6000^3}{3 \times 205 \times 10^3 \times 12\,452 \times 10^4}$$
$$= 17.57 - 8.09 = 9.48 \, \text{mm}.$$

$\delta/\text{height} = 9.48/6000 = 1/632 < 1/300$ (Table 8).
The column is satisfactory with respect to deflection.

7.9 Crane columns

7.9.1 Types

Three common types of crane columns used in single-storey industrial buildings are shown in Figure 7.33. These are:

(1) a column of uniform section carrying the crane beam on a bracket;
(2) a laced crane column;
(3) a compound column fabricated from two universal beams or built up from
    plate.

Only the design of a uniform column used for light cranes will be discussed
here. Types (2) and (3) are used for heavy cranes.

7.9.2 Loading

A building frame carrying a crane is shown in Figure 7.34(a). The hook load is
placed as far as possible to the left to give the maximum load on the column.
The building, crane and wind loads are shown in the figure in (b), (c) and (d),
respectively.

7.9.3 Frame action and analysis

In order to determine the values of moments in the columns the frame as a
whole must be considered. Consider the frame shown in Figure 7.34(a), where
the columns are of uniform section pinned at the top and fixed at the base. The
separate load cases are discussed.

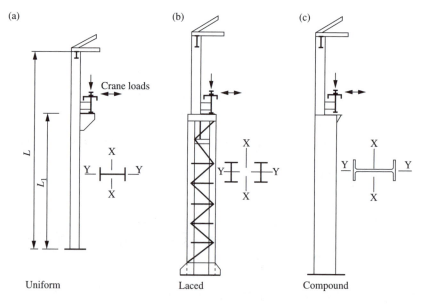

Figure 7.33 Types of crane columns

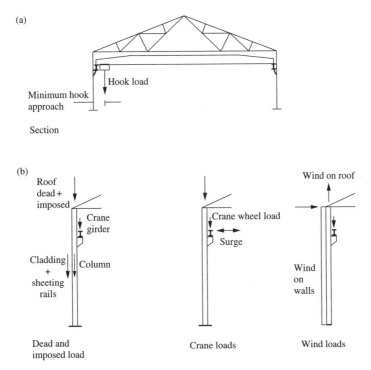

(a)

Hook load

Minimum hook
approach

Section

(b)

Roof
dead +
imposed

Crane
girder

Cladding
+
sheeting
rails

Column

Dead and
imposed load

Crane wheel load

Surge

Crane loads

Wind on roof

Wind
on
walls

Wind loads

Figure 7.34 Loads on crane columns

(1) Dead and imposed load

The dead and imposed loads from the roof and walls are taken as acting axially
on the column. The dead load from the crane girder causes moments as well
as axial load in the column. (See the crane wheel load case below.)

(2) Vertical crane wheel loads

The vertical crane wheel loads cause moments as well as axial load in each
column. The moments applied to each column are unequal, so the frame sways
(as shown in Figure 7.35(a)) and a force P is transmitted through the bottom
chord.

Consider the column ABC in Figure 7.35(a). The deflection at the top is calculated for the moment from the crane wheel loads M_1, and force P separately
using the moment area method. The separate moment diagrams are shown in
Figure 7.35(b).

The deflection due to M_1 is:

$$\delta = M_1 L_1 (L - L_1/2)/EI$$

where, L is the column height, L_1 the height to the crane rail and EI the
column rigidity.

The deflection due to the load P at the column top is

$$\delta_2 = PL^3/3EI$$

(a)

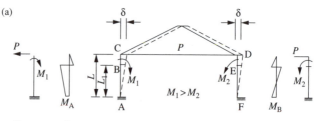

Frame are column moments

(b)

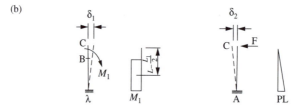

Moments and load causing deflection in column ABC

(c)

Column moments due to crane surge

Figure 7.35 Vertical and horizontal crane-wheel loads and moments

The frame deflection is

$$\delta = \delta_1 - \delta_2$$

Equating deflections at the top of each column gives:

$$\frac{M_1 L_1}{EI}\left(L - \frac{L_1}{2}\right) - \frac{PL^3}{3EI} = \frac{M_2 L_1}{EI}\left(L - \frac{L_1}{2}\right) + \frac{PL^3}{3EI}$$

where $P = 3L_1(L - L_1/2)\,(M_1 + M_2)/2L^3$.
The moments in the column can now be calculated.

If the self-weight of the crane girder applies a moment M to each column then the force in the bottom chord is:

$$P_1 = \frac{L_1 M}{L_3}(L - L_1/2).$$

(3) Crane surge

In Figure 7.35(c), the crane surge load S is the same each side and each column acts as a free cantilever. The loads and moments for this case are shown in the figure.

(4) Wind loads

Wind loads on this type of frame were treated in Section 7.8.1.

(5) Load combinations

The separate load combinations and load factors γ_f to be used in design are given in Table 2 of BS 5950: Part 1. The load cases and load factors are:

(1) 1.4 Dead + 1.6 Imposed + 1.6 Vertical Crane Load,
(2) 1.4 Dead + 1.6 Imposed + 1.6 Horizontal Crane Load,
(3) 1.4 Dead + 1.6 Imposed + 1.4 (Vertical and Horizontal Crane Loads),
(4) 1.2 (Dead + Imposed + Wind + Vertical and Horizontal Crane Loads).

It may not be necessary to examine all cases. Note that in case (2) there is no impact allowance on the vertical crane wheel loads.

7.9.4 Design procedure

(1) Effective lengths for axial compression

The effective lengths for axial compression for a uniform column carrying the crane girder on a bracket are given in Appendix D of BS 5950: Part 1.
    In Figure 7.33(a), the effective lengths are:

x–x axis: $L_E = 1.5\,L$
y–y axis: $L_E = 0.85\,L$,

The code specifies that the crane girder must be held in position longitudinally by bracing in the braced bays. If the base is not fixed in the y–y direction, $L_E = 1.0\,L_1$.

(2) Effective length for calculating the buckling resistance moment

The reader should refer to Sections 4.3.5, 4.3.6 and Table 14 of BS 5950: Part 1. The crane girder forms an intermediate restraint to the cantilever column. Section 4.3.6 states in this case that the member is to be treated as a beam between restraints and Table 14 is to be used to determine the effective length L_E. A value of $L_E = 0.85\,L_1$, may be used for this case.

(3) Column design

The column is checked for local capacity at the base and overall buckling.

(4) Deflection

The deflection limitation for columns in single-storey buildings applies. In Section 2.5.1 and Table 5 of BS 5950: Part 1, the limit for the column top is height/300. However, the code also states that in the case of crane surge and wind only the greater effect of either need be considered in any load combination.

7.9.5 Example: design of a crane column

(1) Building frame and loading

The single-storey building frame shown in Figure 7.36(a) carries a 50 kN electric overhead crane. The frames are at 5 m centres and the length of the building is 30 m. The static crane wheel loads are shown in Figure 7.36(b). The crane beams are simply supported, spanning 5 m between columns, and the weight of a beam is approximately 4 kN. The arrangement of the column and crane beam with the end clearance and eccentricity are shown in Figure 7.36(c).

Dead and imposed loads

The roof and wall loads are the same as for the building in Section 7.8.3. The loads are:

Dead loads
Roof   = 24.23 kN
Walls   = 10.5 kN
Crane column + bracket = 4.0 kN
Crane beam   = 4.0 kN

Dead load at column base = 42.73 kN

Dead load at crane girder level = 27.86 kN
Imposed load – Roof   = 37.5 kN

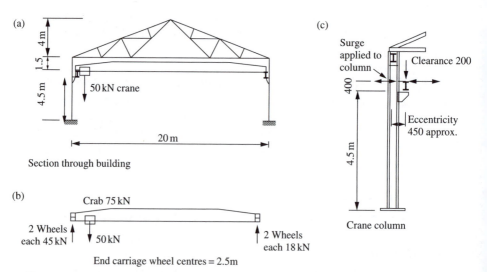

(a)

4 m
1.5
4.5 m

50 kN crane

20 m

Section through building

(b)

Crab 75 kN

2 Wheels
each 45 kN   ▼ 50 kN   2 Wheels
                                          each 18 kN
End carriage wheel centres = 2.5m

Crane

(c)

Surge
applied to
column   Clearance 200
400

Eccentricity
450 approx.

4.5 m

Crane column

Figure 7.36 Building frame with crane

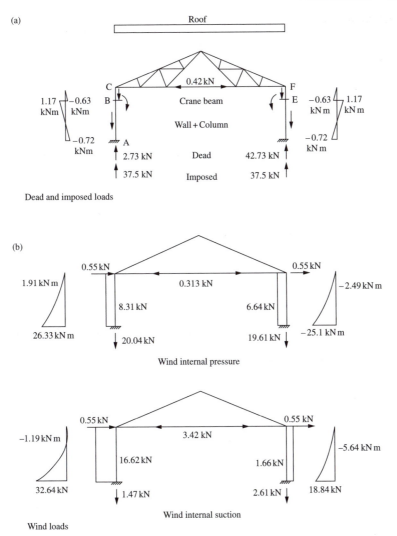

Figure 7.37 Crane column loads and moments

The eccentric dead load of the crane beams cause small moments in the columns.

In Figure 7.37(a), the applied moment to each column:

$$M = 4 \times 0.45 = 1.8 \, \text{kN m}.$$

The load P_1 in the bottom chord of the truss (see above):

$$P_1 = \frac{3 \times 4.5 \times 1.8}{6^3}\left(6 - \frac{4.5}{2}\right) = 0.42 \, \text{kN}$$

Column moments:

$$3M_{BC} = 0.42 \times 1.5 = -0.63 \, \text{kN m},$$
$$M_{BA} = 1.8 - 0.63 = 1.17 \, \text{kN m},$$
$$M_{AB} = 1.8 - (0.42 \times 6) = -0.72 \, \text{kN m}.$$

Wind loads

The wind loads are the same as for the building in Section 7.8.3 and wind load and column moments are shown in Figure 7.37(b).

Vertical crane wheel loads

The crane wheel loads, including impact, are:

Maximum wheel loads $= 45 + 25\% = 56.25 \, \text{kN}$,
Light side-wheel loads $= 18 + 25\% = 22.5 \, \text{kN}$.

To determine the maximum column reaction, the wheel loads are placed equidistant about the column, as shown in Figure 7.38(a). The column reaction and moment for the maximum wheel loads are:

$$R_1 = 2 \times 56.25 \times 3.75/5 = 84.375 \, \text{kN},$$
$$M_1 = 84.3 \times 0.45 = 37.87 \, \text{kN m}.$$

For the light side-wheel loads:

$$R_2 = 2 \times 22.5 \times 3.75/5 = 33.75 \, \text{kN},$$
$$M_2 = 33.75 \times 0.45 = 15.19 \, \text{kN}.$$

The load in the bottom chord is (see above):

$$P_2 = \frac{3 \times 4.5}{2 \times 6^3} \left(6 - \frac{4.5}{2} \right) (37.87 + 15.19) = 6.22 \, \text{kN}$$

The moments for column ABC are:

$$M_{BC} = -6.22 \times 1.5 = -9.33 \, \text{kN m},$$
$$M_{BA} = 37.87 - 9.33 = 28.54 \, \text{kN m},$$
$$M_A = 37.87 - (6.22 \times 6) = 0.55 \, \text{kN m}.$$

(a)

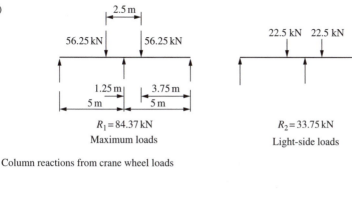

Column reactions from crane wheel loads

(b)

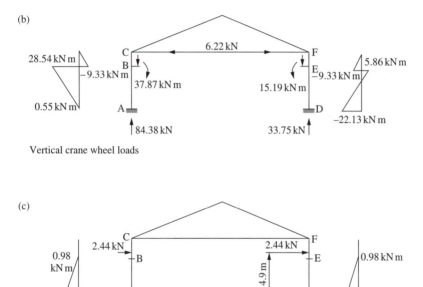

Vertical crane wheel loads

(c)

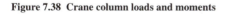

Horizontal crane wheel loads

Figure 7.38 Crane column loads and moments

These moments and the moments for column DEF are shown in Figure 7.38(b).

Crane surge loads

The horizontal surge load per wheel

$$= 0.1(50 + 15)/4 = 1.63 \text{ kN}.$$

The column reaction from surge loads

$$R_3 = 2 \times 1.63 \times 3.75/5 = 2.45 \text{ kN}.$$

The moments at the column base are

$$M_A = 2.45 \times 4.9 = 12.0 \, kN \, m.$$

The loads and moments are shown in Figure 7.38(c).

(2) Design load combinations

Consider column ABC with wind internal suction case, maximum crane wheel loads and crane surge. The design loads and moments for three load combinations are:

(1) Dead + Imposed + Vertical Crane Loads
    Base

$$P = (1.4 \times 42.73) + (1.6 \times 37.5) + (1.6 \times 84.38) = 254.83 \, kN$$
$$M = (1.4 \times 0.72) + (1.6 \times 0.55) = -0.13 \, kN \, m$$

Crane girder

$$P = (1.4 \times 27.86) + (1.6 \times 37.5) + (1.6 \times 84.38) = 234.01 \, kN$$
$$M = (1.4 \times 1.17) + (1.6 \times 28.54) = 47.3 \, kN \, m$$

(2) Dead + Imposed + Vertical and Horizontal Crane Loads
    Base

$$P = (1.4 \times 42.73) + (1.6 \times 37.5) + (1.4 \times 84.38) = 237.95 \, kN$$
$$M = -(1.4 \times 0.72) + 1.4(0.55 - 12) = -17.04 \, kN \, m$$

Crane girder

$$P = (1.4 \times 27.86) + (1.6 \times 37.5) + (1.4 \times 84.38) = 217.14 \, kN$$
$$M = (1.4 \times 1.17) \times (28.54 + 0.98) = 42.97 \, kN \, m$$

(3) Dead + Imposed + Wind Internal Suction + Vertical and Horizontal Crane Loads
    Base

$$P = 1.2[42.73 + 37.5 - 1.47 + 84.38] = 195.77 \, kN$$
$$M = 1.2[-0.72 + 32.64 + 0.55 + 12.0] = 53.36 \, kN \, m$$

Crane girder

$$M = 1.2[1.17 - 1.19 + 28.54 + 0.98] = 35.4 \, kN \, m$$

Note that design conditions arising from notional horizontal loads specified in Clause 2.4.2.3 of BS 5950: Part 1 are not as severe as those in condition (3) above.

(3) Column design

Try 406×140 UB 46, the properties of which are:

$$A = 59.0\,\text{cm}^2; \quad S_x = 888.4\,\text{cm}^3; \quad Z_x = 777.8\,\text{cm}^3;$$

$$r_x = 16.29\,\text{cm}, \quad r_y = 3.02\,\text{cm}, \quad x = 38.8; \quad I_x = 15\,647\,\text{cm}^4.$$

Local capacity check at base

The reader should refer to the example in Section 7.8.3. The section can be shown to be plastic: design strength $p_y = 275\,\text{N/mm}^2$:

$$M_{cx} = 275 \times 888.4/10^3 = 244.3\,\text{kN m}$$

Interaction expression:

Case (1): $\dfrac{254.83 \times 10}{59.0 \times 275}$ + Moment negligible $= 0.157 < 1.0$

Case (2): $\dfrac{237.95 \times 10}{590.275} + \dfrac{17.04}{244.3} = 0.217 < 1.0$

Case (3): $\dfrac{195.77 \times 10}{59.0 \times 275} + \dfrac{53.36}{244.3} = 0.339 < 1.0$

The section is satisfactory. Case (3) is the most severe load combination. Compressive strength:

$$\lambda_x = 1.5 \times 6000/162.9 = 55.25,$$
$$\lambda_y = 0.85 \times 4500/30.2 = 112.6,$$
$$p_c = 106.4\,\text{N/mm}^2 \quad \text{(Table 24(c))}.$$

Bucking resistance moment:

$$L_E = 0.85 \times 4500 = 3825\,\text{mm},$$
$$\lambda = 3825/30.2 = 112.6,$$
$$x = 38.8,$$
$$\eta = 1.0 \text{ (conservative estimate)},$$
$$P_b = 138.2\,\text{N/mm}^2 \quad \text{(Table 16)},$$
$$M_b = 128.2 \times 888.4/10^3 = 122.7\,\text{kN m}.$$

Interaction expression:

$$\frac{195.77 \times 10}{59.0 \times 106.4} + \frac{53.36}{122.7} = 0.74 < 1.0$$

The column is satisfactory.

Deflection at column cap

The reader should refer to Figures 7.37 and 7.38:
Deflection due to crane surge δ_s:

$$EI\,\delta_s = 12 \times 10^6 \times 4900 \times 4367/2 = 1.284 \times 10^{14}.$$

Deflection due to wind, δ_w

$$EI\,\delta_w = 16620 \times 6000^3/8 - 2.87 \times 10^3 \times 6000^3/3 = 2.421 \times 10^{14}.$$

Add deflection from crane wheel loads to that caused by wind load:

$$EI\,\delta = 2.421 \times 10^{14} + 37.87 \times 10^6 \times 4500 \times 3750$$
$$- 6220 \times 6000 \times 10^3/3$$
$$= 4.334 \times 10^{14}$$
$$\delta = \frac{4.334 \times 10^{14}}{205 \times 10^3 \times 15\,647 \times 10^4} = 13.51\,\text{mm},$$
$$\delta/\text{height} = 13.51/6000 = 1/444 < 1/300.$$

The deflection controls the column size.

7.10 Column bases

7.10.1 Types and loads

Column bases transmit axial load, horizontal load and moment from the steel column to the concrete foundation. The main function of the base is to distribute the loads safely to the weaker material.
The main types of bases used are shown in Figure 7.39. These are:

(1) slab base;
(2) gusseted base; and
(3) pocket base.

With respect to slab and gusseted bases, depending on the values of axial load and moment, there may be compression over the whole base or compression over part of the base and tension in the holding-down bolts. Bases subjected to moments about the major axis only are considered here. Horizontal loads are resisted by shear in the weld between column and base plates, shear in the holding-down bolts and friction and bond between the base and the concrete. The horizontal loads are generally small.

7.10.2 Design strengths

(1) Base plates

The design strength of the plate p_{yp} is given in Section 4.13.2.2 of BS 5950: Part 1. This is to be taken from Table 9 but is not to exceed 275 N/mm^2.

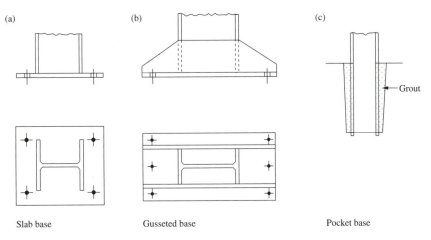

(a)   (b)   (c)

Grout

Slab base   Gusseted base   Pocket base

Figure 7.39 Column bases

(2) Holding-down bolts

The strengths of bolts are given in Table 34 of the code (see Sections 10.2.2 and 10.2.3). The tensile stress area should be used in the design check for bolts in tension.

(3) Concrete

The column base is set on steel packing plates and grouted in. Mortar cube strengths vary from 25 to 40 N/mm^2. The bearing strength is given in Section 4.13.1 of the code as $0.4 f_{cu}$, where f_{cu}, is the cube strength at 28 days. For design of pocket bases, the compressive strength of the structural concrete is taken from BS 8110: Part 1.

7.10.3 Axially loaded slab base

(1) Code requirements and theory

This type of base is used extensively with thick steel slabs being required for heavily loaded columns. The slab base is free from pockets where corrosion may start and maintenance is simpler than with gusseted bases.

The design of slab bases with concentric loads is given in Section 4.13.2.2 of BS 5950: Part 1. This states that where the rectangular plate is loaded by an I, H, channel, box or rectangular hollow section its minimum thickness should be:

$$t_p = k \left(\frac{3.0w}{p_{yp}} \right)^{0.5}$$

but not less than the flange thickness of the column supported, where k is the largest perpendicular distance from the edge of the effective portion of the base plate to the face of the column cross-section, T the flange thickness of the column, w the pressure on the underside of the base assuming uniform distribution and p_{yp} the design strength of the base plate.

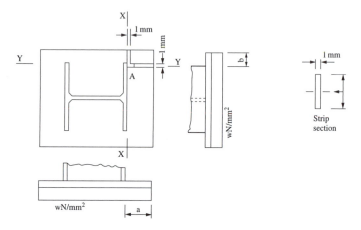

Figure 7.40 Column base plate moments

(2) Weld: column to slab

The code states in Section 4.13.3 that where the slab and column end are in tight contact the load is transmitted in direct bearing. The surfaces in contact would be machined in this case. The weld only holds the base slab in position. Where the surfaces are not suitable to transmit the load in direct bearing the weld must be designed to transmit the load.

7.10.4 Axially loaded slab base: example

A column consisting of a 305 × 305 UC 198 carries an axial dead load of 1600 kN and an imposed load of 800 kN. Adopting a square slab, determine the size and thickness required. The cube strength of the concrete grout is 25 N/mm^2. Use Grade S275 steel.

Design load $= (1.4 \times 1600) + (1.6 \times 800) = 3520\,\text{kN}.$

Effective area method

Required area of base plate, $A_{\text{reg}} = F_{\text{c}}/0.6 f_{\text{cu}}$

$$= 3520 \times 10^3/(0.6 \times 25) = 234\,667\,\text{mm}^2$$

Provide 600 × 600 mm^2 square base plate:

The area provided, $A_{\text{provide}} = 360\,000\,\text{mm}^2 > A_{\text{reg}}.$

The arrangement of the column on the base plate is shown in Figure 7.41, from this the effective area is:

$$A_{\text{eff}} = (D + 2c)(B + 2c) - (D - 2c - 2T)(B - t)$$
$$= (DB + 2Bc + 2Dc + 4c^2)$$
$$\quad - (DB - 2Bc - 2BT - DT - Dt + 2tc + 2Tt)$$
$$= 4c^2 + (2D + 4B - 2t)c + (2BT + Dt - 2Tt)$$

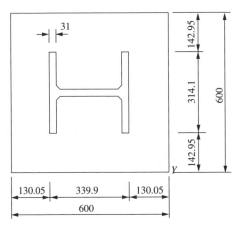

Figure 7.41 Axial loaded base: example

where D and B are the depth and width of the universal steel column used, T and t are the flange and web thicknesses of the UC and c is the perpendicular spread distance as defined in Clause 4.13.2.2 of BS 5950.

The effective area is approximately equal to:

$$A_{\text{eff}} = 4c^2 + c(\text{sectional perimeter}) + \text{sectional area} = A_{\text{reg}}.$$

Now, sectional perimeter $= 1.87 \times 10^3$ mm
sectional area $= 25\,200\,\text{mm}^2$
Equating yields $4c^2 + (1.87 \times 10^3)c + (25.2 \times 10^3) = 234\,667$
Solving the equation, $c = 93.38$.
Check to ensure no overlapping occur:

$$D + 2c = 339.9 + 2 \times 93.38 = 526\,\text{mm} < D_{\text{p}} = 600\,\text{mm (ok)}$$
$$B + 2c = 314.1 + 2 \times 93.38 = 501\,\text{mm} < B_{\text{p}} = 600\,\text{mm. (ok)}$$

The thickness of the base plate is given by:

$$t_{\text{p}} = c \left(\frac{3.0w}{p_{\text{yp}}} \right)^{0.5}$$

Assume that the thickness of plate is less than 40 mm. Design strength for S275 plate, $p_{\text{yp}} = 265\,\text{N/mm}^2$ (Table 9).

$$t_{\text{p}} = 93.38 \left(\frac{3 \times 0.6 \times 25}{265} \right)^{0.5} = 38.38\,\text{mm}.$$

Provide 40 mm thickness of base plate is adequate.

The column flange thickness is 31.4 mm. Make the base plate 40 mm thick. Use 6 mm fillet weld all round to hold the base plate in place. The surfaces are to be machined for direct bearing. The holding-down bolts are nominal but four No. 24 mm diameter bolts would be provided.

Base slab: $600 \times 600 \times 40\,\text{mm}^3$ thick.

Problems

7.1 A Grade S275 steel column having 6.0 m effective length for both axes is to carry pure axial loads from the floor above. If a 254 × 254 UB 89 is available, check the ultimate load that can be imposed on the column. The self weight of the column may be neglected.

7.2 A column has an effective length of 5.0 m and is required to carry an ultimate axial load of 250 kN, including allowance for self-weight. Design the column using the following sections:

(1) universal column section;
(2) circular hollow section;
(3) rectangular hollow section.

7.3 A column carrying a floor load is shown in Figure 7.42. The column can be considered as pinned at the top and the base. Near the mid-height it is propped by a strut about the minor axis. The column section provided is an 457 × 152 UB 60 of Grade 43 steel. Neglecting its self-weight, what is the maximum ultimate load the column can carry from the floor above?

7.4 A universal beam, 305 × 165 UB54, is cased in accordance with the provisions of Clause 4.14.1 of BS 5950. The effective length of the column is 6.0 m. Check that the section can carry an axial load of 720 kN.

7.5 A Grade S275 steel 457 × 152 UB 60 used as a column is subjected to uniaxial bending about its major axis. The design data are as follows:

Ultimate axial compression = 1000 kN,
Ultimate moment at top of column = +200 kN m,
Ultimate moment at bottom of column = −100 kN m,
Effective length of column = 7.0 m.

Determine the adequacy of the steel section.

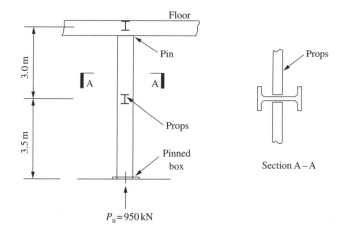

Figure 7.42

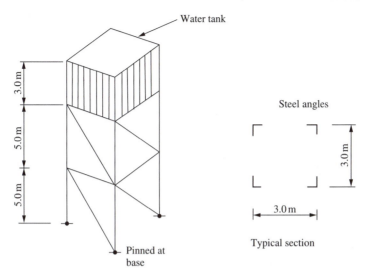

Water tank

Steel angles

3.0 m

5.0 m

5.0 m

3.0 m

3.0 m

Pinned at
base

Typical section

Figure 7.43

7.6 A column between floors of a multi-storey building frame is subjected to biaxial bending at the top and bottom. The column member consists of a Grade S275 steel 305×305 UC158 section. Investigate its adequacy if the design load data are as follows:

Ultimate axial compression $= 2300$ kN,
Ultimate moments,
Top-about major axis $= 300$ kN m,
  about minor axis $= 50$ kN m,
Bottom-about major axis $= 150$ kN m,
  about minor axis $= -80$ kN m.
Effective length of column $= 6.0$ m.

7.7 A steel tower supports a water tank of size $3 \times 3 \times 3$ m^3. The self-weight of the tank is 50 kN when empty. The arrangement of the structure is shown in Figure 7.43. Other design data are given below:

Unit weight of water $= 9.81$ kN/m^3,
Design wind pressure $= 1.0$ kN/m^2.

Use Grade S275 steel angles for all members. Design the steel tower structure and prepare the steel drawings.

8

Trusses and bracing

8.1 Trusses—types, uses and truss members

8.1.1 Types and uses of trusses

Trusses and lattice girders are framed elements resisting in-plane loading by axial forces in either tension or compression in the individual members. They are beam elements, but their use gives a large weight saving when compared with a universal beam designed for the same conditions.

The main uses for trusses and lattice girders in buildings are to support roofs and floors and carry wind loads. Pitched trusses are used for roofs while parallel chord lattice girders carry flat roofs or floors.

Some typical trusses and lattice girders are shown in Figure 8.1(a). Trusses in buildings are used for spans of about 10–50 m. The spacing is usually about 5–8 m. The panel length may be made to suit the sheeting or decking used and the purlin spacing adopted. Purlins need not be located at the nodes but this introduces bending into the chord. The panel spacing is usually 1.5–4 m.

8.1.2 Truss members

Truss and lattice girder members are shown in Figure 8.1(b). The most common members used are single and double angles, tees, channels and the structural hollow sections. I and H and compound and built-up members are used in heavy trusses.

8.2 Loads on trusses

The main types of loads on trusses are dead, imposed and wind loads. These are shown in Figure 8.1(a).

8.2.1 Dead loads

The dead load is due to sheeting or decking, insulation, felt, ceiling if provided, weight of purlins and self weight. This load may range from 0.3 to 1.0 kN/m^2. Typical values are used in the worked examples here.

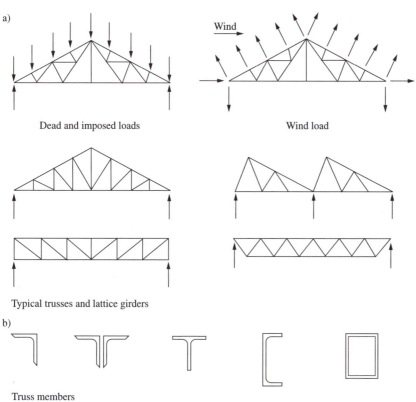

a)

Wind

Dead and imposed loads   Wind load

Typical trusses and lattice girders

b)

Truss members

Figure 8.1 Roof trusses and lattice girders

8.2.2 Imposed loads

The imposed load on roofs is taken from BS 6399: Part 1, Section 6. This loading may be summarized as follows:

(1) Where there is only access to the roof for maintenance and repair— $0.75 \, \text{kN/m}^2$;
(2) Where there is access in addition to that in (1)—$1.5 \, \text{kN/m}^2$.

8.2.3 Wind loads

Readers should refer wind loading to the new BS 6399: Part 2: 'Code of practice for wind loads'. The wind load depends on the building dimensions and roof slope, among other factors. The wind blowing over the roof causes a suction or pressure on the windward slope and a suction on the leeward one (see Figure 8.1(a)). The loads act normal to the roof surface.

Wind loads are important in the design of light roofs where the suction can cause reversal of load in truss members. For example, a light angle member is satisfactory when used as a tie but buckles readily when required to act as a strut. In the case of flat roofs with heavy decking, the wind uplift will not be greater than the dead load, and it need not be considered in the design.

8.2.4 Application of loads

The loading is applied to the truss through the purlins. The value depends on the roof area supported by the purlin. The purlin load may be at a node point, as shown in Figure 8.1(a), or between nodes, as discussed in Section 8.3.3 below. The weight of the truss is included in the purlin point loads.

8.3 Analysis of trusses

8.3.1 Statically determinate trusses

Trusses may be simply supported or continuous, statically determinate or redundant and pin jointed or rigid jointed. However, the most commonly used truss or lattice girder is single span, simply supported and statically determinate. The joints are assumed to be pinned, though, as will be seen in actual construction, continuous members are used for the chords. This assumption for truss analysis is stated in Section 4.10 of BS 5950: Part 1.

A pin-jointed truss is statically determinate when:

$$m = 2j - 3$$

where m denotes the number of members and j the number of joints.

8.3.2 Load applied at the nodes of the truss

When the purlins are located at the node points the following manual methods of analysis are used:

(1) Force diagram—This is the quickest method for pitched roof trusses.
(2) Joint resolution—This is the best method for a parallel chord lattice girder.
(3) Method of sections—This method is useful where it is necessary to find the force in only a few members.

The force diagram method is used for the analysis of the truss in Section 8.6. An example of use of the method of sections would be for a light lattice girder where only the force in the maximum loaded member would be found. The member is designed for this force and made uniform throughout.

A matrix analysis program can be used for truss analysis. In Chapter 11, the roof truss for an industrial building is analysed using a computer program.

8.3.3 Loads not applied at the nodes of the truss

The case where the purlins are not located at the nodes of the truss is shown in Figure 8.2(a). In this case the analysis is made in two parts:

(1) The total load is distributed to the nodes as shown in Figure 8.2(b). The truss is analysed to give the axial loads in the members.

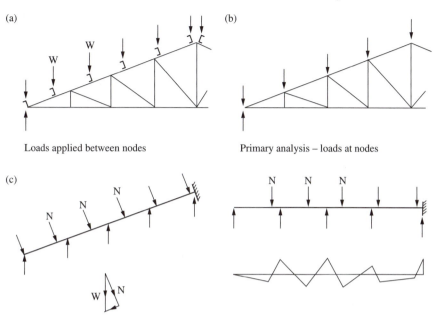

(a)

W

W

Loads applied between nodes

(b)

Primary analysis – loads at nodes

(c)

N

N

N

N

N

N

W

N

Secondary analysis of top chord as a continuous beam

Figure 8.2 Loads applies between nodes of truss

(2) The top chord is now analysed as a continuous beam loaded with the normal component of the purlin loads as shown in Figure 8.2(c). The continuous beam is taken as fixed at the ridge and simply supported at the eaves. The beam supports are the internal truss members. The beam is analysed by moment distribution. The top chord is then designed for axial load and moment.

8.3.4 Eccentricity at connections

BS 5950: Part 1 states in Section 6.1.2 that members meeting at a joint should be arranged so that their centroidal axes meet at a point. For bolted connections, the bolt gauge lines can be used in place of the centroidal axes. If the joint is constructed with eccentricity, then the members and fasteners must be designed to resist the moment that arises. The moment at the joint is divided between the members in proportion to their stiffness.

8.3.5 Rigid-jointed trusses

Moments arising from rigid joints are important in trusses with short, thick members. BS 5950: Part 1 states in Section 4.10 that secondary stresses from these moments will not be significant if the slenderness of chord members in the plane of the truss is greater than 50 and that of most web members is greater than 100. Rigid jointed trusses may be analysed using a matrix stiffness analysis program.

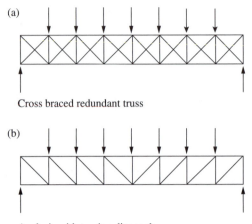

(a)

Cross braced redundant truss

(b)

Analysis with tension diagonals

Figure 8.3 Cross-brace lattice girder

8.3.6 Deflection of trusses

The deflection of a pin-jointed truss can be calculated using the strain energy method. The deflection at a given node is:

$$\delta = \sum PuL/AE,$$

where P is the force in a truss member due to the applied loads, u the force in a truss member due to unit load applied at the node and in the direction of the required deflection, L the length of a truss member, A the area of a truss member and E the Young's modulus.

A computer analysis gives the deflection as part of the output.

8.3.7 Redundant and cross-braced trusses

A cross-braced wind girder is shown in Figure 8.3. To analyse this as a redundant truss, it would be necessary to use the computer analysis program.

However, it is usual to neglect the compression diagonals and assume that the panel shear is taken by the tension diagonals, as shown in Figure 8.3(b). This idealization is used in design of cross bracing (see Section 8.7).

8.4 Design of truss members

8.4.1 Design conditions

The member loads from the separate load cases—dead, imposed and wind—must be combined and factored to give the critical conditions for design. Critical conditions often arise through reversal of load due to wind, as discussed below. Moments must be taken into account if loads are applied between the truss nodes.

8.4.2 Struts

(1) Maximum slenderness ratios

Maximum slenderness ratios are given in Section 4.7.3.3 of BS 5950: Part 1. For lightly loaded members these limits often control the size of members. The maximum ratios are:

(1) Members resisting other than wind load—180,
(2) Members resisting self-weight and wind load—250,
(3) Members normally acting as ties but subject to reversal of stress due to wind—350.

The code also states that the deflection due to self-weight should be checked for members whose slenderness exceeds 180. If the deflection exceeds the ratio length/1000, the effect of bending should be taken into account in design.

(2) Limiting proportions of angle struts

To prevent local buckling, limiting width/thickness ratios for single angles and double angles with components separated are given in Table 11 of BS 5950: Part 1. These are shown in Figure 8.4.

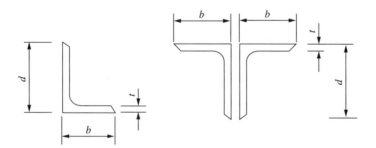

Section	Limiting proportions	
	b/t and $d/t \leqslant$	$(b+d)/t \leqslant$
Plastic	$9.0\,\varepsilon$	–
Compact	$10.0\,\varepsilon$	–
Semi-compact	$15.0\,\varepsilon$	$24\,\varepsilon$

$$\varepsilon = (275/p_y)^{0.5}$$

Figure 8.4 Limiting proportions for single and double-angle struts

(3) Effective lengths for compression chords

The compression chord of a truss or lattice girder is usually a continuous member over a number of panels or, in many cases, its entire length. The chord is supported in its plane by the internal truss members and by purlins at right angles to the plane as shown in Figure 8.5.

The code defines the length of chord members in Section 4.10 as:

(1) In the plane of the truss-panel length—L_1,
(2) Out of the plane of the truss-purlin spacing—L_2.

The rules from Section 4.7.2 of the code can then be used to determine the effective lengths. The slenderness ratios for single and double angle chords are shown in Figure 8.5. Note that truss joints reduce the in-plane value of effective length.

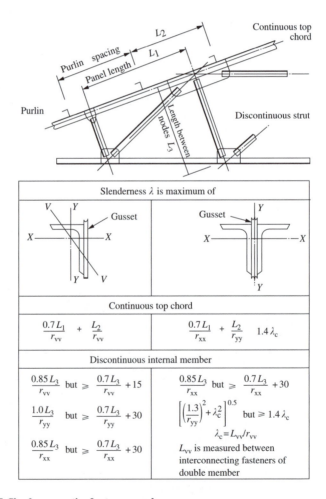

Figure 8.5 Slenderness ratios for truss members

(4) Effective lengths of discontinuous internal truss members

Discontinuous internal truss members, a single angle or double angle connected to a gusset at each end, are shown in Figure 8.5. The effective lengths for the cases where the connections contain at least two fasteners or the equivalent in welding are given in Section 4.7.10 and Table 25 of the code. The slenderness ratios are shown in the figure. The length L_3 is the distance between truss nodes. The effective lengths for other sections are given in the code.

(5) Design procedure

The code states in Section 4.7.10.1 that the end eccentricity for discontinuous struts may be ignored and the design made for an axially loaded member.

For single angles or double angles with members separated, the compression resistance for plastic, compact or semi-compact sections is given by:

$$P_c = A_g p_c$$

where A_g denotes gross area.

From Table 23 in the code, the strut curve (Table 24(c)) is selected to obtain the compressive strength P_c.

If the section is slender, the effective area method is used to determine the member compressive strength, refer to Section 7.4.9 for the method and example.

8.4.3 Ties

The effective area is used in the design of discontinuous angle ties. This was set out in Section 6.4 above. Tension chords are continuous throughout all or the greater part of their length. Checks will be required at end connections and splices.

8.4.4 Members subject to reversal of load

In light roofs, the uplift from wind can be greater than the dead load. This causes a reversal of load in all members. The bottom chord is the most seriously attached member and must be supported laterally by a lower chord bracing system, as shown in Figure 1.2. It must be checked for tension due to dead and imposed loads and compression due to wind load.

8.4.5 Chords subjected to axial load and moment

Angle top chords of trusses may be subjected to axial load and moment, as discussed in Section 8.3.3 above. The buckling capacity for axial load is calculated in accordance with Section 8.4.2 (3) above.

Simplified method of calculating the buckling resistance moment for single equal angles is given in clause 4.3.8.3 of the steel code. Subject to bending about the x–x axis and the section ratio $b/t \leq 15\varepsilon$, the buckling moment of resistance is:

When the heel of the angle is in compression,

$$M_b = 0.8 p_y Z_x.$$

When the heel of the angle is in tension,

$$M_b = p_y Z_x (1350\varepsilon - L_E/r_v)/1625\varepsilon$$

where L_E denotes effective length from section 4.3.5 of BS 5950, r_v the least radius of gyration about $v-v$ axis and Z_x the smaller section modules about the $x-x$ axis.

The same expression may be used for a double angle chord as shown in Figure 8.5 using the maximum of L_1/r_x, and L_2/r_y. The design check is elastic, so the value of Z to be used depends on where the check is made. For example, for the hogging moment at a node, the minimum value of Z_x is used. It is conservative not to take moment gradient into account, that is, $m = 1.0$.

The interaction expression for the overall buckling check is:

$$F/(A_g p_c) + M/M_b < 1.0$$

where F = axial load,
A_g = gross area of chord,
p_c = compressive strength,
M = applied moment,
M_b = buckling resistance moment.

8.5 Truss connections

8.5.1 Types

The following types of connections are used in trusses:

(1) Column cap and end connections;
(2) Internal joints in welded construction;
(3) Bolted site joints-internal and external.

The internal joints may be made using a gusset plate or the members may be welded directly together. Some typical connections using gussets are shown in Figure 8.6. In these joints, all welding is carried out in the fabrication shop. The site joints are bolted.

8.5.2 Joint design

Joint design consists of designing the bolts, welds and gusset plate.

(1) Bolted joints

The load in the member is assumed to be divided equally between the bolts. The bolts are designed for direct shear and the eccentricity between the bolt gauge line and the centroidal axis is neglected (see Figure 8.7(a)). The bolts and gusset plate are checked for bearing.

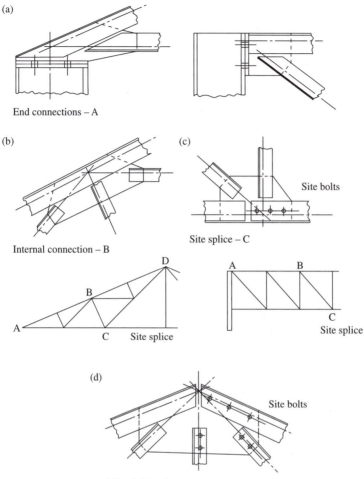

Figure 8.6 Truss and lattice girder connections

(2) Welded joints

In Figure 8.7(a), the weld groups can be balanced as shown. That is, the centroid of the weld group is arranged to coincide with the centroidal axis of the angle in the plane of the gusset. The weld is designed for direct shear.

If the angle is welded all round, the weld is loaded eccentrically, as shown in Figure 8.7(b). However, the eccentricity is generally not considered in practical design because much more weld is provided than is needed to carry the load.

(3) Gusset plate

The gusset plate transfers loads between members. The thickness is usually selected from experience but it should be at least equal to that of the members to be connected.

(a)

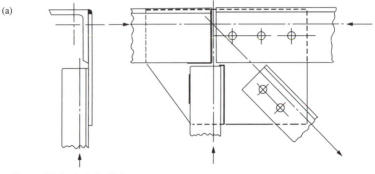

Shop welded - site bolted joint

(b)

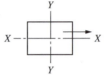

Eccentrically loaded weld group

(c)

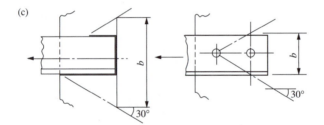

Effective width of gusset plate

Figure 8.7 Truss connections and gusset plate design

The actual stress conditions in the gusset are complex. The direct stress in the plate can be checked at the end of the member assuming that the load is dispersed at 30° as shown in Figure 8.7(c). The direct load on the width of dispersal b should not exceed the design strength of the gusset plate. In joints where members are close together, it may not be possible to disperse the load. In this case a width of gusset equal to the member width is taken for the check.

8.6 Design of a roof truss for an industrial building

8.6.1 Specification

A section through an industrial building is shown in Figure 8.8(a). The frames are at 5 m centres and the length of the building is 45 m. The purlin spacing on

(a)

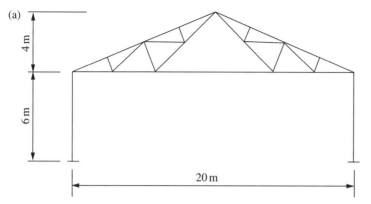

Section through building

(b)

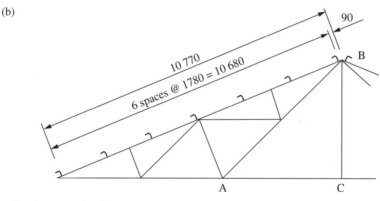

Arrangement of purlins

Figure 8.8 A pitched-roof truss

the roof is shown in Figure 8.8(b). The loading on the roof is as follows:

(1) Dead load—measured on the slope length
     Sheeting and insulation board $= 0.25\,\mathrm{kN/m^2}$,
     Purlins $= 0.1\,\mathrm{kN/m^2}$,
     Truss $= 0.1\,\mathrm{kN/m^2}$,
     Total dead load $= 0.45\,\mathrm{kN/m^2}$.
(2) Imposed load measured on plan $= 0.75\,\mathrm{kN/m^2}$,
     Imposed load measured on slope $= 0.75 \times 10/10.77 = 0.7\,\mathrm{kN/m^2}$.

Design the roof truss using angle members and gusseted joints. The truss is to be fabricated in two parts for transport to site. Bolted site joints are to be provided at A, B and C as shown in the figure.

8.6.2 Truss loads

(1) Dead and imposed loads

Because of symmetry only one half of the truss is considered.

Deadloads :
    End panelpoints $= 1/8 \times 0.45 \times 10.77 \times 5 = 3.03\,\text{kN}$,
    Internal panelpoints $= 2 \times 3.03 = 6.06\,\text{kN}$.
Imposedloads :
    End panelpoint $= 3.03 \times 0.7/0.45 = 4.71\,\text{kN}$,
    Internal panelpoints $= 2 \times 4.71 = 9.42\,\text{kN}$.

The dead loads are shown in Figure 8.10.

(2) Wind loads

The effective wind speed is taken as $V_e = 33.3\,\text{m/s}$.
Dynamic pressure $q = 0.613 \times 33.3^2/10^3 = 0.68\,\text{kN/m}^2$.

The external pressure coefficients C_{pe} from the internal pressure coefficients C_{pi} are shown in Figure 8.9. The values used are where there is only a negligible probability of a dominant opening occurring during a severe storm. C_{pi} is taken as the more onerous of the values $+0.2$ or -0.3.

For the design of the roof truss, the condition of maximum uplift is the only one that need be investigated. A truss is selected from Section FH of the roof shown in Figure 8.9(a), where C_{pe} is a maximum and C_{pi} is taken as $+0.2$, the case of internal pressure. The wind load normal to the roof is:

$$0.68\,(C_{pe} - C_{pi}).$$

The wind loads on the roof are shown in Figure 8.9(c) for the two cases of wind transverse and longitudinal to the building.

The wind loads at the panel points normal to the top chord for the case of wind longitudinal to the building are:

End panel points $= 1/8 \times 0.612 \times 10.77 \times 5 = 4.12\,\text{kN}$,
Internal panel points $= 8.24\,\text{kN}$.

The wind loads are shown in Figure 8.11.

8.6.3 Truss analysis

(1) Primary forces its truss members

Because of symmetry of loading in each case, only one half of the truss is considered. The truss is analysed by the force diagram method and the analyses are shown in Figures 8.10 and 8.11. Note that members 4–5 and 5–6 must be replaced by the fictitious member 6–X to locate point 6 on the force diagram. Then point 6 is used to find points 4 and 5.

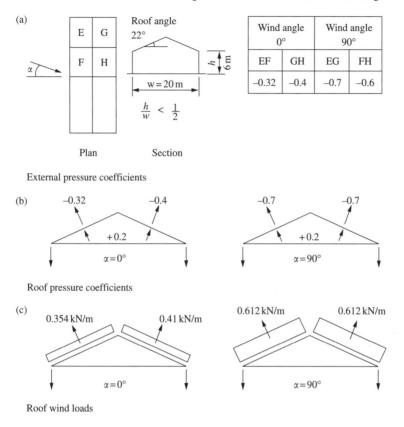

Figure 8.9 Wind loads on the roof truss

The dead load case is analysed and the forces due to the imposed loads are found by proportion. The case for maximum wind uplift is analysed. The forces in the members of the truss are tabulated for dead, imposed and wind load in Table 8.1.

(2) Moments in the top chord

The top chord is analysed as a continuous beam for moments caused by the normal component of the purlin load from dead load:

Purlin load $= 1.78 \times 5 \times 0.35 = 3.12\,\text{kN}$,
Normal component $= 3.12 \times 10/10.77 = 2.89\,\text{kN}$,
End purlin $L = 2.89 \times 0.98/1.78 = 1.59\,\text{kN}$.

The top chord loading is shown in Figure 8.12(a). The fixed end moments are:

Span AB
$$M_A = 0,$$
$$M_{BA} = 2.89 \times 1.78(2.69^2 - 1.78^2)/2 \times 2.69^2 = 1.44\,\text{kN m}.$$

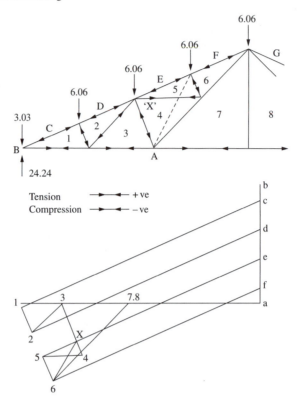

Figure 8.10 Dead-load analysis

Span BC

$$M_{BC} = 2.89[(0.87 \times 1.82^3) + (2.65 \times 0.04^2)]/2.69^2 = 1.15\,\text{kN m},$$
$$M_{CB} = 2.89[(1.82 \times 0.87^2) + (0.04 \times 2.65^2)]/2.69^2 = 0.66\,\text{kN m}.$$

Span CD

$$M_{CD} = 2.89 \times 1.74 \times 0.95^2/2.69^2 = 0.63\,\text{kN m},$$
$$M_{DC} = 2.89 \times 0.95 \times 1.74^2/2.69^2 = 1.15\,\text{kN m}.$$

Span DE

$$M_{DE} = 2.89 \times 0.82 \times 1.87^2/2.69^2 + 1.59 \times 2.6 \times 0.09^2/2.69^2 = 1.15\,\text{kN m}$$
$$M_{ED} = 2.89 \times 1.87 \times 0.82^2/2.69^2 + 1.59 \times 2.6 \times 0.09^2/2.69^2 = 0.51\,\text{kN m}$$

The distribution factors are:

Joint B BA:BC = 0.75:1 = 0.43:0.57,
Joints C and D = 0.5:0.5.

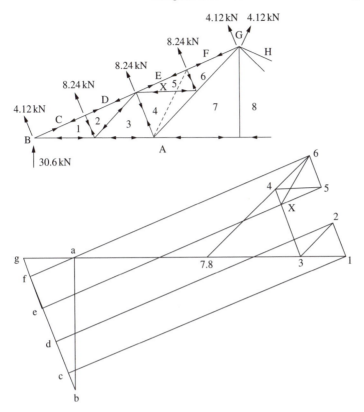

Figure 8.11 Wind-load analysis

The moment distribution is shown in Figure 8.12(b) and the reactions and internal moments for the separate spans are shown in Figure 8.13(a). The bending-moment diagram for the complete top chord is shown in Figure 8.13(b) and the axial loads from the force diagram in Figure 8.13(c).

8.6.4 Design of truss members

Part of the member design depends on the joint arrangement. The joint detail and design are included with the member design. Full calculations for joint design are not given in all cases.

(1) Top chord

The top chord is to be a continuous member with a site joint at the ridge.

Member C-1 at eaves

The maximum design conditions are at B (Figure 8.13).

Dead + imposed load:
  Compression $F = -(1.4 \times 57.1) - (1.6 \times 88.8) = -222.0\,\text{kN}$.
  Moment $M_B = -(1.4 \times 1.32) - (1.6 \times 1.32 \times 0.7/0.35) = -6.07\,\text{kN m}$.

Table 8.1 Forces in members of roof truss (kN)

Member	Dead load	Imposed load	Wind load
Top chord			
C-1	−57.1	−88.8	72.1
D-2	−54.9	−85.4	72.1
E-5	−52.6	−81.8	74.9
F-6	−50.3	−78.2	74.9
Bottom chord			
A-1	53.9	82.4	−66.9
A-3	45.5	−70.8	−55.9
A-7	30.3	47.1	−32.1
Struts			
1–2	−5.6	−8.7	8.2
3–4	−11.3	−17.6	17.6
5–6	−5.6	−8.7	8.2
Ties			
2–3	7.6	11.8	−11.1
4–5	7.6	11.8	−14.1
4–7	15.2	23.6	−23.7
6–7	22.7	35.3	−34.8
7–8	0.0	0.0	0.0

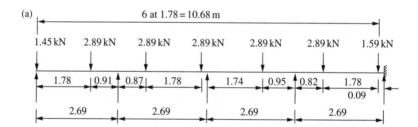

(a)

6 at 1.78 = 10.68 m

1.45 kN   2.89 kN   2.89 kN   2.89 kN   2.89 kN   2.89 kN   1.59 kN

1.78   0.91   0.87   1.78   1.74   0.95   0.82   1.78   0.09

2.69   2.69   2.69   2.69

Top chord dead loads

(b)

	0.43	0.57		0.5	0.5		0.5	0.5	
0.0	1.44	−1.15		0.66	−0.63		1.15	−1.15	0.63
	−0.12	−0.17		−0.02	−0.02		0.0	0.0	
	0.0	−0.01		−0.09	0.0		0.0	0.0	0.0
	0.0	0.0		0.05	0.05		0.0	0.0	
0.0	1.32	−1.32		0.6	−0.6		1.15	−1.15	0.63

Moment distribution, top chord analysis

Figure 8.12 Top-chord analysis

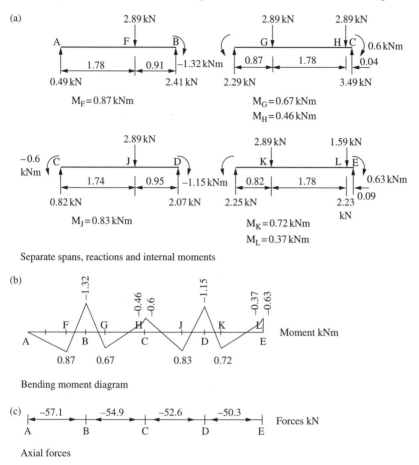

Separate spans, reactions and internal moments

Bending moment diagram

Axial forces

Figure 8.13 Top-chord moments and forces

Dead + wind load:
   Tension $F = -57.1 + (1.4 \times 72.1) = 43.8$ kN.

Try 2 No. $100 \times 100 \times 10$ angles with 10 mm thick gusset plate. The gross section is shown in Figure 8.14(a). The propertiesare:

$$A = 19.2 \times 2 = 38.4\,\text{cm}^2; \quad r_x = 3.05\,\text{cm}, \quad r_y = 3.05\,\text{cm};$$
$$Z_x = 24.8 \times 2 = 49.6\,\text{cm}^3.$$

Design strength $p_y = 275\,\text{N/mm}^2$ (Table 9).
   Check the section classification using Table 11.

$$d/t = 100/10 = 10 < 15,$$
$$(b + d)/t \times 200/10 = 20 < 24.$$

The section is semi-compact.

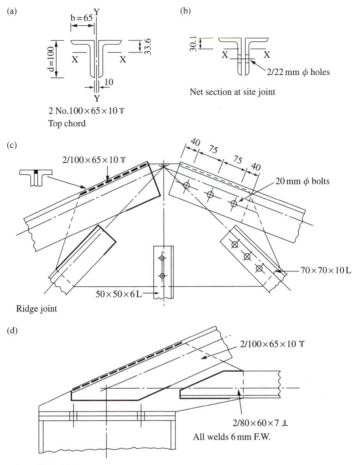

Figure 8.14 Top-chord design details

The slenderness is the maximum of:

$L_E/r_x = 0.7 \times 2690/30.5 = 61.8$.
$L_E/r_y = 1780/30.5 = 58.4$.

Compressive strength $p_c = 197\,\text{N/mm}^2$ (Table 24(c)).

The buckling resistance moment for a single angle is given in Section 4.3.8 of the code. The two angles act together and the slenderness is less than 100. Thus:

$$M_b = 0.8 \times 275 \times 49.6/10^3 = 10.9\,\text{kN m}.$$

Interaction expression for overall buckling:

$$\frac{222.02 \times 10}{38.4 \times 197} + \frac{6.07}{10.9} = 0.86 < 1.0.$$

Provide 2 No. $100 \times 100 \times 10$ angles.

The member will be satisfactory for the case of dead plus wind load.

Member F-6 at ridge

The ridge joint is shown in Figure 8.14(c). A bolted site joint is provided in the chord on one side. The design conditions are:

Compression $F = (1.4 \times 50.3) + (1.6 \times 78.2) = 195.54\,\text{kN}$,
Moment $M_E = -(1.4 \times 0.57) - (1.6 \times 0.57 \times 0.07/0.35) = 2.62\,\text{kN m}$.

The net section is shown in Figure 8.14(b):

Net area $= 34.0\,\text{cm}^2$; $Z_x = 40.69\,\text{cm}^3$ (minimum).
$M_b = 0.8 \times 275 \times 40.69/101 = 8.95\,\text{kN m}$.

Interaction expression for local capacity is:

$$\frac{195.54 \times 10}{275 \times 34.0} + \frac{2.62}{8.95} = 0.50 < 1.0.$$

The section is satisfactory.

Eaves joint (see Figure 8.14 (d))

The member is connected to both sides of the gusset so the gross section is effective in resisting load (see Section 4.6.3.3 of the code).

Compression $F = 222.02\,\text{kN}$,
Length of 6 mm fillet, strength 0.9 kN/mm, required:
$\quad = 222.02/0.9 = 246.6\,\text{mm}$.

This may be balanced around the member as shown in the figure. More weld has been provided than needed.

The bearing capacity of the gusset is checked at the end of the member on a width of 100 mm. No dispersal of the load is considered because of the compact arrangement of the joint:

Bearing capacity $= 275 \times 100 \times 10/10^3 = 275\,\text{kN}$.

The gusset is satisfactory.

Ridge joint (see Figure 8.14(c))

Try three No. 20 mm diameter Grade 4.6 bolts at the centres shown on the figure. From Table 4.2 in the code, the double shear value is 76.2 kN. The bolts resist:

Direct shear $= 195.54\,\text{kN}$,
Moment $= 2.62\,\text{kN}$,
Direct shear per bolt $= 195.54/3 = 65.18\,\text{kN}$,
Shear due to moment $= 2.62/0.15 = 17.47\,\text{kN}$,
Resultant shear $= (65.18^2 + 17.47^2)^{0.5} = 67.4\,\text{kN}$.

The bolts are satisfactory.

The shop-welded joint must be designed for moment and shear. The weld is shown in the figure. The gusset will be satisfactory.

(2) Bottom chord

The bottom chord is to have two site joints at P and R, as shown in Figure 8.15.

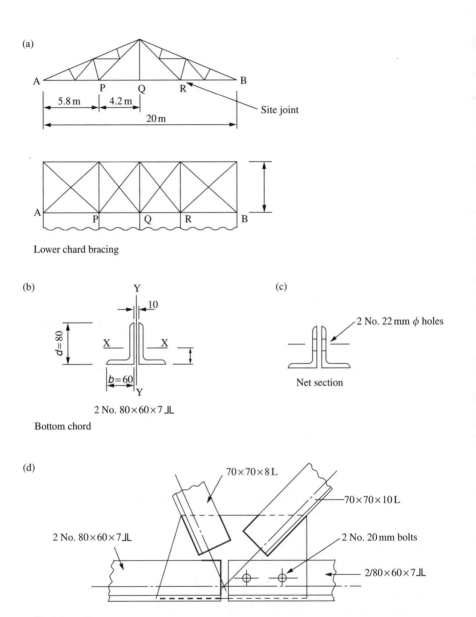

Figure 8.15 Bottom-chord design details

Member A-1

The design conditions are:

Dead + imposed load:
   Tension $F = (1.4 \times 53) + (1.6 \times 82.4) = 206.04\,\text{kN}$.

Dead + wind load:
   Compression $F = 53 - (1.4 \times 66.9) = -40.66\,\text{kN}$.

Try two No. $80 \times 60 \times 7$ angles. The section is shown in Figure 8.15(b). The properties are:

$$A = 18.76\,\text{cm}^2; \quad r_x = 2.51\,\text{cm}; \quad r_y = 2.34\,\text{cm}.$$

The section is semi-compact.
   When the bottom chord is in compression due to uplift from wind, lateral supports will be provided at P, Q and R by the lower chord bracing shown in Figure 8.15(a). The effective length for buckling about the $y–y$ axis is 5800 mm:

$L_E/r_y = 5800/23.4 = 248$,
$p_c = 28.4\,\text{N/mm}^2$ (Table 24(c)).
$P_c = 18.76 \times 28.4/10 = 53.2\,\text{kN}$.

At end A, the angles are connected to both sides of the gusset:

$P_t = 275 \times 18.76/10 = 515.9\,\text{kN}$.

Provide 2 No. $80 \times 60 \times 7$ angles. The wind load controls the design.
   The connection to the gusset is shown in Figure 8.14(d). The length of 6 mm weld required $= 206.04/0.9 = 228.9\,\text{mm}$.
The weld is placed as shown in the figure. The gusset is satisfactory.

Member A-7

Dead + imposed load:
   Tension $F = (1.4 \times 30.3) + (1.6 \times 47.1) = 117.8\,\text{kN}$.
Dead + wind load:
   Compression $= 30.3 - (1.4 \times 32.1) = -14.64\,\text{kN}$.

Member A-7 is connected to the gusset by 2 No. 20 mm diameter bolts in double shear as shown in Figure 8.15(d):

Net area $= 18.76 - 2 \times 22 \times 7/10^2 = 15.68\,\text{cm}^2$,
$P_t = 15.68 \times 275/10 = 431.2\,\text{kN}$.

The member will also be satisfactory when acting in compression due to wind. The shear capacity of the joint $= 2 \times 76.2 = 152.4\,\text{kN} > 117.8\,\text{kN}$.

(3) Internal members

Members 4–7, 6–7

Design for the maximum load in members 6–7:

Dead + imposed load:
   Tension $F = (1.4 \times 22.7) + (1.6 \times 35.3) = 88.26\,\text{kN}$.

(a)

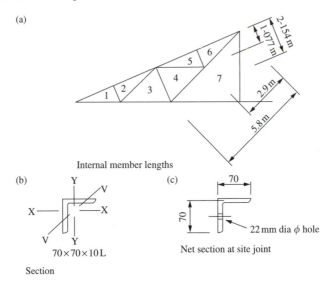

Internal member lengths

(b)

X ——— X

V   Y
$70 \times 70 \times 10\,L$

Section

(c)

22 mm dia ϕ hole

Net section at site joint

Figure 8.16 Design for members 4–7, 6–7 and 3–4

Dead + wind load:
  Compression load = $22.7 - (1.4 \times 34.8) = -26.02\,kN$.

Try $70 \times 70 \times 10$ angle. The member lengths and section are shown in Figure 8.16. The properties are:

$A = 13.1\,cm^2$; $r_x = 2.09\,cm$; $r_v = 1.36\,cm$.

The slenderness values are calculated below (see Figure 8.5).

  Members 6–7 buckling about the v–v axis:

    $\lambda = 0.85 \times 2900/13.6 = 181.2$ but
    $\geq 0.7 \times 2900/13.6 + 15 = 164.3$,
    $\lambda = 2900/20.9 = 138.8$ but $\geq 0.7 \times 2900/20.9 + 30 = 127.1$.

  Members 6–7 and 4–7 buckling laterally:

    $\lambda = 5800/20.9 = 277.5$ but
    $\geq 0.7 \times 5800/20.9 + 30 = 224.2$.
    Compressive strength for $\lambda = 277.5$: $p_c = 23.2\,N/mm^2$ (Table 24(c)).
    Compression resistance: $P_c = 23.2 \times 13.1/10 = 30.4\,kN$.

This is satisfactory.
  The end of member 6–7 is connected at the ridge by bolts, as shown in Figure 8.14(c). The net section is shown in Figure 8.16(c):

Net area of connected leg = $10(65 - 22) = 430\,mm^2$.
Area of unconnected leg = $10 \times 65 = 650\,mm^2$.

Effective area $A_e = 430 + 650 = 1080\,mm^2$.

Tension capacity $P_t = p_y(A_e - 0.5a_2)$
$$= 275 \times (1080 - 0.5 \times 650)/10^3$$
$$= 207.6\,\text{kN} > 88.3\,\text{kN}.$$

This is satisfactory.

Joints for Members 4–7, 6–7

Use 20 mm bolts in clearance holes.

Single shear value $= 38.1\,\text{kN}$ (Table 4.2).
Number of bolts required $= 88.26/38.1 = 2.31$. Use three bolts.

The bolts are shown in Figure 8.14(c) and the welded connection is also shown on the figure. The connection to the site joint at P is shown in Figure 8.15(d).

Members 3–4

The design loads are:

Dead + imposed load:
   Compression $F = -(1.4 \times 11.3) - (1.6 \times 17.6) = -43.98\,\text{kN}.$

Dead + wind load:
   Tension $F = -11.3 + (1.4 \times 17.6) = 13.34\,\text{kN}.$

Try $70 \times 70 \times 8$ angle. The properties are:

$$A = 10.6\,\text{cm}^2; \quad r_x = 2.11\,\text{cm}; \quad r_v = 1.36\,\text{cm}.$$

See Figure 8.16.

$\lambda = 0.85 \times 2154/13.6 = 134.6 \quad$ but $\quad > 0.7 \times 2154/13.6 + 15 = 125.9$
$\quad = 2154/21.1 = 102.1 \quad$ but $\quad > 0.7 \times 2154/21.1 + 30 = 101.5.$
$p_c = 81.4\,\text{N/mm}^2.$
$P_c = 81.4 \times 10.6/10 = 86.2\,\text{kN}.$

A smaller angle could be used but this section will be adopted for uniformity.

Other internal members

All other members are to be $50 \times 50 \times 6$ angles. The design for these members is not given.

(4) Truss arrangement

A drawing of the truss is shown in Figure 8.17 and details for the main joints are shown in Figures 8.14 and 8.15.

8.7 Bracing

8.7.1 General considerations

Bracing is required to resist horizontal loading in buildings designed to the simple design method. The bracing also generally stabilizes the building and ensures that the framing is square. It consists of the diagonal members between

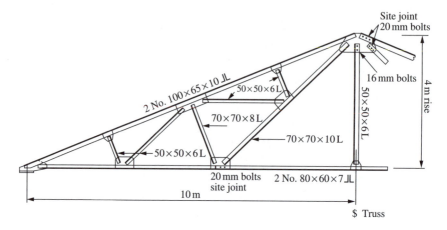

Figure 8.17 Arrangement drawing of truss

columns and trusses and is usually placed in the end bays. The bracing carries the load by forming lattice girders with the building members.

8.7.2 Bracing for single-storey industrial buildings

The bracing for a single-storey building is shown in Figure 8.18(a). The internal frames resist the transverse wind load by bending in the cantilever columns. However, the gable frame can be braced to resist this load, as shown. The wind blowing longitudinally causes pressure and suction forces on the windward and leeward gables and wind drag on the roof and walls. These forces are resisted by the roof and wall bracing shown.

If the building contains a crane an additional load due to the longitudinal crane surge has to be taken on the wall bracing. A bracing system for this case is shown in Figure 8.18(b).

8.7.3 Bracing for a multi-storey building

The bracing for a multi-storey building is shown in Figure 8.19. Vertical bracing is required on all elevations to stabilize the building. The wind loads arc applied at floor level. The floor slabs transmit loads on the internal columns to the vertical lattice girders in the end bays. If the building frame and cladding are erected before the floors are constructed, floor bracing must be provided as shown in Figure 8.19(c). Floor bracing is also required if precast slabs not effectively tied together are used.

8.7.4 Design of bracing members

The bracing can be single diagonal members or cross members. The loading is generally due to wind or crane surge and is reversible. Single bracing members must be designed to carry loads in tension and compression. With cross-bracing, only the members in tension are assumed to be effective and those in compression are ignored.

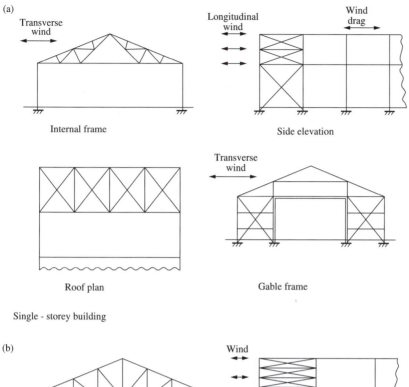

(a)

Single - storey building

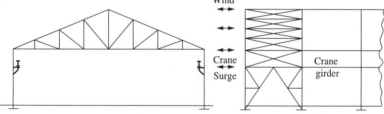

Building with a crane

Figure 8.18 Bracing for single-storey building

The bracing members are the diagonal web members of the lattice girder formed with the main building members, the column, building truss chords, purlins, eaves and ridge members, floor beams and roof joists. The forces in the bracing members are found by analysing the lattice girder. The members are designed as ties or struts, as set out in Section 8.4 above. Bracing members are often very lightly loaded and minimum-size sections are chosen for practical reasons.

8.7.5 Example: bracing for a single-storey building

The gable frame and bracing in the end bay of a single-storey industrial building are shown in Figure 8.20. The end bay at the other end of the building is also braced. The length of the building is 50 m and the truss and column frames are at 5 m centres. Other building dimensions are shown in the figure. Design

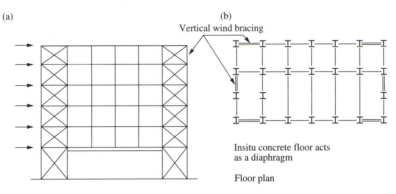

Figure 8.19 Bracing for multi-storey building

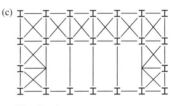

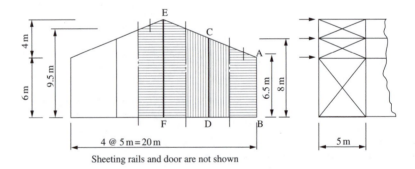

Sheeting rails and door are not shown

Figure 8.20 Gable frame and bracing in end bay

the roof and wall bracing to resist the longitudinal wind loading using Grade S275 steel.

(1) Wind load (refer to CP3: Chapter V: Part 2)

Effective design wind speed $= 33.3\,\text{m/s}$,

Dynamic pressure $q = 0.613 \times 33.3^2/10^3 = 0.68\,\text{kN/m}^2$.

The pressure coefficients on the end walls from are set out in Figure 8.21(a). The overall pressure coefficient from Figure 12 of the code for $D/H < 1.0$ is

(a)

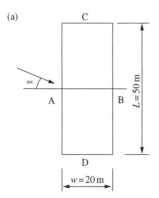

Wind angle	C_{pe} for surface	
∝	C	D
0°	−0.6	−0.6
90°	+0.7	−0.1

External pressure coefficients - end walls

(b)

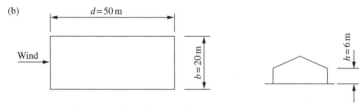

Dimensions for calculating frictional drag

Figure 8.21 Data for calculating wind loads

$C_{pe} = 0.85$. This is not affected by the internal pressure:

Wind pressure $= 0.85 \times 0.68 = 0.578 \, \text{kN/m}^2$.

The method for calculating the frictional drag is given in Clause 2.4.5 of the code:

Cladding—corrugated plastic coated steel sheet.
Factor $C_{f'} = 0.02$ for surfaces with corrugations across the wind direction.
Refer to Figure 8.21(b).
$d/h = 50/6 = 8.33 > 4$,
$d/b = 50/20 = 2.5 < 4$.

The ratio d/h is greater than 4, so the drag must be evaluated.
For $h < b$, the frictional drag is given by:

F' roof $= C_{f'}q \, bd(d - 4h)$
$\qquad = 0.02 \times 0.68 \times 20(50 - 24) = 7.1 \, \text{kN}$.
F' walls $= C_{f'}q \, 2h(d - 4h)$
$\qquad = 0.02 \times 0.68 \times 12(50 - 24) = 4.24 \, \text{kN}$.

The total load is the sum of the wind load on the gable ends and the frictional drag. The load is divided equally between the bracing at each end of the building.

(2) Loads on bracing (see Figure 9.20)

Point E

Load from the wind on the end gable column EF
$= 9.5 \times 5 \times 0.5 \times 0.578 = 13.73\,\text{kN}$,
Reaction at E, top of the column $= 6.86\,\text{kN}$,
Load at E from wind drag on the roof
$= 0.5 \times 7.1 \times 0.25 = 0.9\,\text{kN}$,
Total load at $E = 7.76\,\text{kN}$.

Point C

Load from the wind on the end gable column CD
$= 8 \times 5 \times 0.5 \times 0.578 = 11.56\,\text{kN}$,
Reaction at C, top of the column $= 5.78\,\text{kN}$,
Load C from wind drag on the roof $= 0.9\,\text{kN}$,
Total load at $C = 6.68\,\text{kN}$.

Point A

Load from the wind on building column AB
$= 2.5 \times 6.5 \times 0.5 \times 0.578 = 4.69\,\text{kN}$,
Reaction at A, top of the column $= 2.35\,\text{kN}$,
Load at A from wind drag on roof and wall
$= (0.125 \times 7.1 \times 0.5) + (0.5 \times 4.24 \times 0.25) = 0.97\,\text{kN}$,
Total load at $A = 3.32\,\text{kN}$.

(3) Roof bracing

The loading on the lattice girder formed by the bracing and roof members and the forces in the bracing members are shown in Figure 8.22(a). Note that the members of the cross bracing in compression have not been shown. Forces are transmitted through the purlins in this case. The maximum loaded member is AH:

Design load $= 1.4 \times 15.5 = 20.62\,\text{kN}$.
Try $50 \times 50 \times 6$ angle with 2 No. 16 mm diameter Grade 4.6 bolts in the
    end connections.
Bolt capacity $= 2 \times 25.1 = 50.2\,\text{kN}$ (Table 4.2).
Referring to Figure 8.22(a):
    Net area $A_e = (47 - 18)6 + 282 = 456\,\text{mm}^2$,
    Unconnected area $a_2 = 282\,\text{mm}^2$.
Tension capacity $P_t = p_y(A_e - 0.5a_2)$
$= 275 \times (456 - 0.5 \times 282)/10^3 = 86.6\,\text{kN}$.
Make all the members the same section.

(4) Wall bracing

The load on the wall bracing and the force in the bracing member are shown in Figure 8.22(b).

Design load $= 1.4 \times 21.7 = 30.4\,\text{kN}$.
Provide $50 \times 50 \times 6$ angle.

(a)

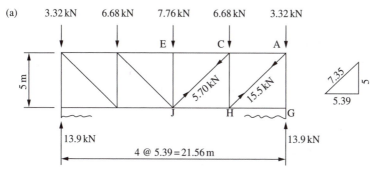

Forces in bracing members

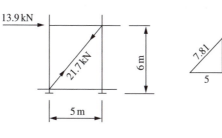

50×50×6 L
19 mm φ hole

Bracing member

Roof bracing

(b)

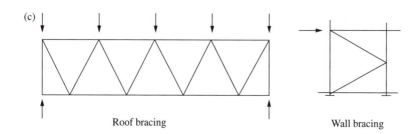

Wall bracing

(c)

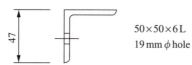

Roof bracing

Wall bracing

Single bracing system

Figure 8.22 Bracing and member forces

(5) *Further considerations*

Design for load on one gable

In a long building, the bracing should be designed for the maximum load at one end. This is the external pressure and internal suction on the gable plus one half of the frictional drag on the roof and 'walls. In this case, if the design is made on this basis:

  Roof bracing: Design load in AH = 25.1 kN,
  Wall bracing: Design load = 39.2 kN.

The 50 × 50 × 6 angles will be satisfactory.

Single bracing system

In the above analysis, cross bracing is provided and the purlins form part of the bracing lattice girder. This arrangement is satisfactory when angle purlins are used. However, if cold-rolled purlins are used the bracing system should be independent of the purlins. A suitable system is shown in Figure 8.22(c), where the roof-bracing members support the gable columns. Circular hollow sections are often used for the bracing members.

8.7.6 Example: bracing for a multi-storey building

The framing plans for an office building are shown in Figure 8.23. The floors and roof are cast *in situ* reinforced concrete slabs which transmit the wind load from the internal columns to the end bracing. Design the wind bracing using Grade S275 steel.

(1) *Wind loads*

The data for the wind loading are:

  The site in country with closest distance to sea upwind 2 km, at the first floor
    the height above ground of 5 m, take:
    Effective design wind speed, $V_e = 26 \, \text{m/s}$.
  Table 4 of BS 6399 part 2, the value of S_b for difference height:
    First floor $H = 5 \, \text{m}$ $S_b = 1.62$,
    Second floor $H = 10 \, \text{m}$ $S_b = 1.78$,
    Roof $H = 15 \, \text{m}$ $S_b = 1.85$.

The design wind speeds and dynamic pressures are:

  First floor $V_e = 26.0 \, \text{m/s}$.
      $q = 0.613 \times 26.0/10^3 = 0.42 \, \text{kN/m}^2$.
  Second floor $V_e = 26.0 \times (1.78/1.62) = 28.5 \, \text{m/s}$.
      $q = 0.49 \, \text{kN/m}^2$.
  Roof level $V_e = 26.0 \times (1.85/1.62) = 29.6 \, \text{m/s}$.
      $q = 0.54 \, \text{kN/m}^2$.

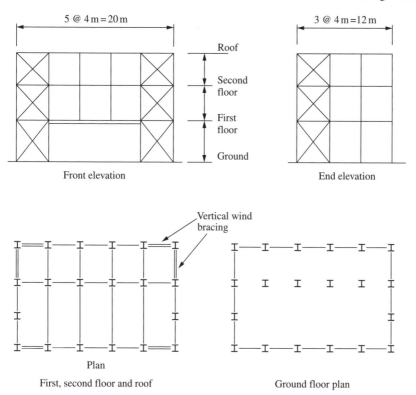

Figure 8.23 Framing plans for a multi-storey building

The force coefficients C_f for wind on the building as a whole are shown in Figure 8.24 for transverse and longitudinal wind:

$$\text{Force} = C_f q A_e,$$

where A_e denotes the effective frontal area under consideration.

Wind drag on the roof and walls need not be taken into account because neither the ratio d/h nor d/h is greater than 4.

(2) Transverse bracing

The loads at the floor levels are:

$$P = 0.981 \times 0.54 \times 2 \times 10 = 10.6 \,\text{kN},$$
$$Q = 0.981 \times 0.49 \times 4 \times 10 = 19.2 \,\text{kN},$$
$$R = 0.981 \times 0.42 \times 4.5 \times 10 = 18.5 \,\text{kN}.$$

The loads are shown in Figure 8.25(a) and the forces in the bracing members in tension are also shown in the figure. The member sizes are selected (see Figure 8.25(c)).

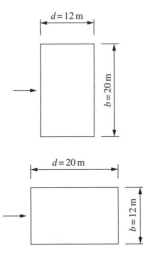

l = length of building = 20 m
w = width of building = 12 m
h = height of building = 13 m

Wind transverse		
l/w	b/d	C_f for h/b=0.65
1.66	1.66	0.981

Wind longitudinal		
l/w	b/d	C_f for h/b=1.08
1.66	0.6	0.816

Figure 8.24 Wind loads: force coefficients

Member QT, RU

Design load $= 1.4 \times 42.1 = 58.9$ kN.

Provide $50 \times 50 \times 6$ angle. The tension capacity allowing for one No. 18 mm diameter holes was calculated as 98.2 kN in Section 8.7.5 (2) above.

Using 16 mm diameter Grade 4.6 bolts:
Capacity 25.1 kN in single shear:
Member QT—provide two bolts each end.
Member RU—provide three bolts each end.

Member S V

Design load $= 1.4 \times 77.3 = 108.22$ kN.

Try $70 \times 70 \times 6$ angle with 20 mm diameter bolts with capacity in single shear of 39.2 kN per bolt.

No. of bolts required at each end = 3.

For the angle
Net area = 538.7 mm^2.
Tension capacity = 148.1 kN.

Note that a $60 \times 60 \times 6$ angle with 16 mm bolts could be used but five bolts would be required in the end connection.

(3) Longitudinal bracing

The loads at the floor levels are:

$A = 0.816 \times 0.54 \times 2 \times 6 = 5.29$ kN,
$B = 0.816 \times 0.49 \times 4 \times 6 = 9.6$ kN,
$C = 0.816 \times 0.42 \times 4.5 \times 6 = 9.25$ kN.

(a)

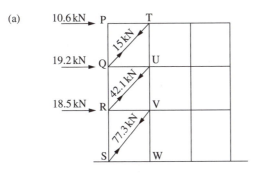

Transverse bracing

(b)

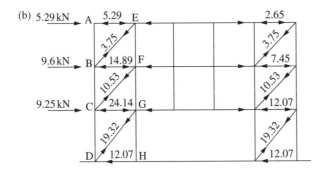

Longitudinal bracing

(c)

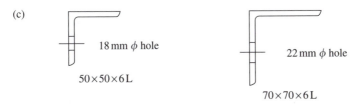

Bracing members

Figure 8.25 Bracing member design

The loads are shown in Figure 8.25(b) and are divided between the bracing at each end of the building. Those in the bracing members in tension are shown in the figure.

The maximum design load for member DG is
$$= 1.4 \times 19.32 = 27.1 \text{ kN}.$$

Provide $50 \times 50 \times 6$ angles for all members. Tension capacity 98.2 kN with one 18 mm diameter hole. Two No. 16 mm diameter bolts are required at the ends of all bracing member.

Problems

8.1 A flat roof building of 18 m span has 1.5 m deep trusses at 4 m centres. The trusses carry purlins at 1.5 m centres. The total dead load is $0.7\,\text{kN/m}^2$ and the imposed load is $0.75\,\text{kN/m}^2$:

(1) Analyse the truss by joint resolution.
(2) Design the truss using angle sections with welded internal joints and bolted field splices.

8.2 A roof truss is shown in Figure 8.26. The trusses are at 6 m centres, the length of the building is 36 m and the height to the eaves is 5 m. The roof loading is:

Dead load = $0.4\,\text{kN/m}^2$ (on slope),
Imposed load = $0.75\,\text{kN/m}^2$ (on plan).

The wind load is to be estimated using BS 6399: Part 2. The building is located on the outskirts of a city and the basic wind speed is 26 m/s.

(1) Analyse the truss for the roof loads.
(2) Analyse the top chord for the loading due to the purlin spacing shown. The dead load from the roof and purlins is $0.32\,\text{kN/m}^2$.
(3) Design the truss.

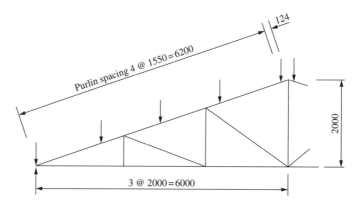

Figure 8.26

8.3 A section through a building is shown in Figure 8.27. The roof trusses are supported on columns at A and B and cantilever out to the front of the building. The front has roller doors running on tracks on the floor. The frames are at 6 m centres and the length of the building is 48 m. The roof load is:

Dead load = $0.45\,\text{kN/m}^2$ (on slope),
Imposed load = $0.75\,\text{kN/m}^2$ (on plan).

The wind loads are to be in accordance with BS 6399: Part 2. The basic wind speed is 26 m/s and the location is in the suburbs of a city. The structure should be analysed for wind load for the two conditions of doors opened and closed. Analyse and design the truss.

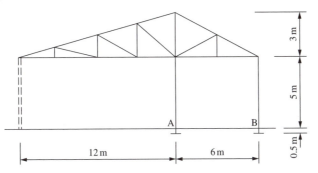

Figure 8.27 Building frame with cantilever truss

8.4 The end framing and bracing for a single-storey building are shown in Figure 8.28. The location of the building is on an industrial estate on the outskirts of a city in the north-east of England. The length of the building is 32 m. The wind loads are to be in accordance with BS 6399: Part 2. Design the bracing.

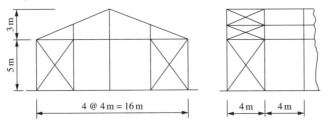

Figure 8.28 Bracing for a factory building

8.5 The framing for a square tower building is shown in Figure 8.29. The bracing is similar on all four faces. The building is located in a city centre in an area where the basic wind speed is 30 m/s. Design the bracing.

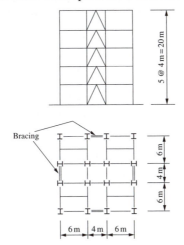

Figure 8.29 Framing for a square-tower building

Portal frames

9.1 Design and construction

9.1.1 Portal type and structural action

The single-storey clear-span building is in constant demand for warehouses, factories and many other purposes. The clear internal appearance makes it much more appealing than a trussed roof building and it also requires less maintenance and heating. The portal may be of three-pinned, pinned-base or fixed-base construction as shown in Figure 9.1(a). The pinned-base portal is the most common type adopted because of the greater economy in foundation design over the fixed-base type.

In plane, the portal resists the following loads by rigid frame action (see Figure 9.1(b)):

(a) dead and imposed loads acting vertically;
(b) wind causing horizontal loads on the walls and generally uplift loads on the roof slopes.

In the longitudinal direction, the building is of simple design and diagonal bracing is provided in the end bays to provide stability and resist wind load on the gable ends and wind friction on sides and roof (see Figure 9.1(d)).

9.1.2 Construction

The main features in modern portal construction shown in Figure 9.1(c) are:

(a) *Columns*—uniform universal beam section;
(b) *Rafters*—universal beams with haunches usually of sections 30–40% lighter than the columns;
(c) *Eaves and ridge connections*—site-bolted joints using grade 8.8 bolts where the haunched ends of the rafters provide the necessary lever arm for design. Local joint stiffening is required;
(d) *Base*—nominally pinned with two or four holding down bolts;
(e) *Purlins and sheeting rails*—cold-formed sections spaced at not greater than 1.75–2 m centres;
(f) *Stays from purlins and rails*—these provide lateral support to the inside flange of portal frame members;

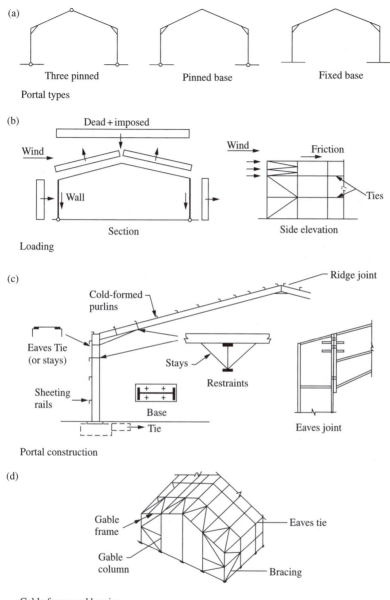

(a)

Three pinned   Pinned base   Fixed base

Portal types

(b)

Dead + imposed

Wind

Wall

Section

Wind   Friction

Ties

Side elevation

Loading

(c)

Ridge joint

Cold-formed purlins

Eaves Tie (or stays)

Stays

Restraints

Sheeting rails

Base

Tie

Eaves joint

Portal construction

(d)

Gable frame

Gable column

Eaves tie

Bracing

Gable frame and bracing

Figure 9.1 Portal frames

(g) *Gable frame*—a braced frame at the gable ends of the buildings;
(h) *Bracing*—provided in the end bay in roof and walls;
(i) *Eaves and ridge ties*—may be provided in larger-span portals, though now replaced by stays from purlins or sheeting rails.

9.1.3 Foundations

The pinned-base portal is generally adopted because it is difficult to ensure fixity without piling and it is more economical to construct. It is also advantageous to provide a tie through the ground slab to resist horizontal thrust due to dead and imposed load as shown in Figure 9.1(c).

9.1.4 Design outline

The code states that either elastic or plastic design may be used. Plastic design gives the more economical solution and is almost universally adopted. The design process for the portal consists of:

(a) analysis—elastic or plastic;
(b) design of members taking into account of flexural and lateral torsional buckling with provision of restraints to limit out-of-plane buckling;
(c) sway stability check in the plane of the portal;
(d) joint design with provision of stiffeners to ensure all parts are capable of transmitting design actions;
(e) serviceability check for deflection at eaves and apex.

Procedures for elastic and plastic design are set out in BS 5950-1: 2000 and are discussed below.

9.2 Elastic design

9.2.1 Code provision

The provisions from BS 5950-1: 2000 are summarized as follows:

(a) Clause 5.2.2: analysis is made using factored loads.
(b) Clause 5.4.1: the capacity and buckling resistance of members are to be checked using Section 4 of the code.
(c) Clause 5.5.2: the stability of the frame should be checked using Section 2.4.2 of the code. This states that, in load combination 1, the notional horizontal loads of 0.5% of the factored dead + imposed load should be applied. These loads are applied at eaves level and act with 1.4× dead load +1.6× imposed load. They are not to be combined with wind loads. In load combinations 2 and 3, the horizontal component of the factored wind load should not be taken as less than 1% of the factored dead load applied horizontally.

9.2.2 Portal analysis

The most convenient manual method of analysis is to use formulae from the *Steel Designers Manual* (see Further reading at the end of this chapter). A general load case can be broken down into separate cases for which solutions are given and then these results are recombined. Computer analysis is the most convenient method to use, particularly for wind loads and load combinations. The output gives design actions and deflections.

   The critical load combination for design is 1.4× dead load +1.6× imposed load. The wind loads mainly cause uplift and the moments are generally in the opposite direction to those caused by the dead and imposed loads. The bending

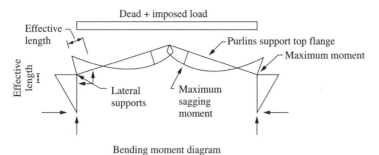

Bending moment diagram

Figure 9.2 Bending moment diagram

moment diagram for the dead and imposed load case is given in Figure 9.2. This shows the inside flange of the column and rafter near the eaves to be in compression and hence the need for lateral restraints in those areas.

As noted in Section 9.1.2, portals are constructed with haunched joints at the eaves. The primary purpose of the haunch is to provide the lever arm to enable the bolted site joints to he made. It is customary to neglect the haunch in elastic analysis. If the haunch is large in comparison with the rafter length, more moment is attracted to the eaves and a more accurate solution is obtained if it is taken into consideration. This can be done by dividing the haunch into a number of parts and using the properties of the centre of each part in a computer analysis.

9.2.3 Effective lengths

Member design depends on the effective lengths of members in the plane and normal to the plane of the portal. Effective lengths control the design strengths for axial load and bending.

(1) In-plane effective lengths

The method for estimating the in-plane effective lengths for portal members was developed by Fraser and is reproduced with his kind permission. Computer programs for matrix stability analysis are also available for determining buckling loads and effective lengths. Only the pinned-base portal with symmetrical uniform loads (that is, the critical load case dead + imposed load on the roof) is considered. Reference should be made to Fraser for unsymmetrical load cases (see further reading at the end of this chapter).

If the roof pitch is less than 10°, the frame may be treated as a rectangular portal and the effective length of the column found from a code limited frame sway chart such as Figure E2 from Appendix E of BS 5950-1: 2000. For pinned-base portals with a roof slope greater than 10° Fraser gives the following equation for obtaining the column effective length which is a close fit to results from matrix stability analyses:

$$k_c = \frac{L_E}{L_C} = 2 + 0.45 G_R$$

where

$$G_R = \frac{L_R I_C}{L_C},$$

L_E is the effective length of column,
L_C and L_R are the actual length of column and rafter, respectively, and
I_C and I_R are second moment of area of column and rafter.

Similar results can be obtained using the limited frame sway chart (Figure E2, BS 5950) if, as Fraser suggests, the pitched roof portal is converted to an equivalent rectangular frame. The beam length is made equal to the total rafter length as shown in Figure 9.3. To apply Figure E2 of the code:

Column top:

$$k_1 = \frac{I_C/L_C}{(I_C/L_C) + (I_R/L_R)}$$

Column base:

$k_2 = 1.0$ for pinned base
$k_2 = 0.5$ for fixed base

Then L_E/L_C for the column is read from Figure E2 of BS 5950-1: 2000.
The effective length of the rafter can be obtained from the effective length of the column because the entire frame collapses as a unit at the critical load, i.e.

Column:

$$P_{CC} = \frac{\pi E I_C}{K_C^2 L_C^2}$$

Rafter:

$$P_{RC} = \frac{\pi^2 E I_R}{K_R^2 (L_R/2)^2}$$

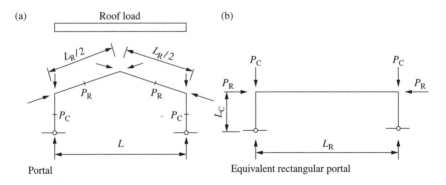

Figure 9.3 Equivalent rectangular portal

hence

$$K_R = K_C \frac{2L_C}{L_R} \sqrt{\frac{P_C I_R}{P_R I_C}}$$

where

E is the Young's modulus, $K_R = L_{ER}/L_R$,
L_{ER} is the Effective length of rafter,
L_R the actual length of rafter,
P_{CC}, P_{RC} the critical loads for column and rafter = load factor × design loads, and
P_C, P_R the average design axial load.

(2) Out-of-plane effective length

The purlins and sheeting rails restrain the outside flange of the portal members. It is essential to provide lateral support to the side flange, i.e. a torsional restraint to both flanges at critical points to prevent out-of-plane buckling. For the pinned-base portal with the bending moment diagram for dead and imposed load shown in Figure 9.2, supports to the inside flange are required at:

(a) eaves (may be stays from sheeting rails or a tie),
(b) within the top half of the column (more than one support may be necessary),
(c) near the point of contraflexure in the rafter (more than one support may be necessary).

The lateral supports and effective lengths about the minor axis for flexural buckling and lateral torsional buckling are shown in Figure 9.2.

9.2.4 Column Design

The column is a uniform member subjected to axial load and moment with moment predominant. The design procedure for the critical load case of dead + imposed load is as follows.

(1) Compressive resistance

Estimate slenderness ratios L_{EX}/r_x, L_{Ey}/r_y
where L_{Ex} and L_{Ey} are the effective lengths for x–x and y–y axis and r_x and r_y are radii of gyration for x–x and y–y axis, respectively.
Compressive strength, p_c from Table 24 of the code
Compressive resistance, $P_c = p_c A_g$.

(2) Moment capacity

For Class 1 and 2 section,

$$M_c = p_y S$$

where p_y is the design strength and S is the plastic modulus.

(3) Buckling resistance moment

The bending moment diagram and effective length are shown in Figure 9.2.
Calculate the equivalent slenderness, λ_{LT}

$$\lambda_{LT} = uv\lambda\sqrt{\beta_w}$$

where

v is from Table 19 of the BS 5950-1: 2000,
u from section table, $\lambda = LE/r_y$ and $\beta_w = 1.0$ for class 1 and 2 section.

Bending strength, p_b from Table 16 of the code
Buckling resistance moment, $M_b = p_b S$.

(4) Interaction expressions

Cross-section capacity

$$\frac{F_c}{A_g p_y} + \frac{m_x M_x}{M_{cx}} \leq 1$$

Member buckling resistance

$$\frac{F_c}{P_c} + \frac{m_x M_x}{p_y Z_x} \leq 1$$

$$\frac{F_c}{P_{cy}} + \frac{m_{LT} M_{LT}}{M_b} \leq 1$$

9.2.5 Rafter design

The rafter is a member haunched at both ends with the moment distribution
shown in Figure 9.2. The portion near the eaves has compression on the inside.
Beyond the point of contraflexure, the inside flange is in tension and is stable.
The top flange is fully restrained by the purlins.

If the haunch length is small, it may conservatively be neglected and
the design made for the maximum moment at the eaves. A torsional
restraint is provided at the first or second purlin point away from the
eaves and the design is made in the same way as for the column. Rafter
design taking the haunch into account is considered under plastic design in
Section 9.3.

9.2.6 Example: Elastic design of a portal frame

(1) Specification

The pinned-base portal for an industrial building is shown in Figure 9.4(a). The
portals are at 5 m centres and the length of the building is 40 m. The building
load are:

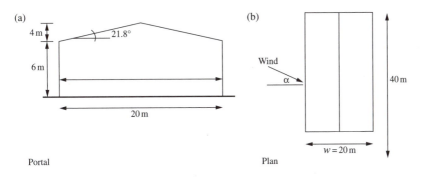

Portal   Plan

Figure 9.4 Pinned base portal

Roof–Dead load measured on slope:

Sheeting	$= 0.10\,\text{kN/m}^2$
Insulation	$= 0.15\,\text{kN/m}^2$
Purlins (Table 4.2, P145/170, 3.97 kg/m at 1.5 m c/c)	$= 0.03\,\text{kN/m}^2$
Rafter 457 × 191 UB 67	$= 0.13\,\text{kN/m}^2$
Total dead load	$= 0.41\,\text{kN/m}^2$
Imposed snow load on plan, BS 6399	$= 0.60\,\text{kN/m}^2$
Imposed services load	$= 0.15\,\text{kN/m}^2$
Total imposed load	$= 0.75\,\text{kN/m}^2$

Wind load—BS 6399: Part 2—Location: Leeds, UK, outskirts of city, 50 m above sea level, 100 m to sea.

Carry out the following work:
   Estimate the building loads.
   Analyse the portal using elastic theory with a uniform section throughout.
      Design the section for the portal using Grade S275 steel.

(2) *Loading*

Roof–Dead	$= 0.41 \times 5 \times 10.77/10$	$= 2.21\,\text{kN/m},$
Imposed	$= 0.75 \times 5$	$= 3.75\,\text{kN/m},$
Design	$= (1.4 \times 2.21) + (1.6 \times 3.75)$	$= 9.09\,\text{kN/m}.$

Notional horizontal load at each column top:

$$= 0.5 \times 0.005 \times 20 \times 9.09 = 0.45\,\text{kN}.$$

Wind – Basic wind speed
   $V_b = 23$ m/s
   $S_a = 1 + (0.001 \times 50) = 1.05$
   $S_d = 1.0$
   $S_s = 1.0$
   $S_p = 1.0$
   Site wind speed, $V_s = V_b \times S_a \times S_d \times S_s \times S_p = 23 \times 1.05 = 24.2$ m/s.

The wind load factors S_b, effective wind speeds V_e, dynamic pressures q, and external pressure coefficients C_{pe} for the portal are shown in Table 9.1. The internal pressure coefficients C_{pi} from Table 16 of the wind code are $+0.2$ or -0.3 for the case where there is a negligible probability of a dominant opening occurring during a severe storm.

Dynamic pressure, $q = 0.613\,V_e^2 \times 10^{-3}\,\text{kN/m}^2$

Wind load, $w = 5q(C_{pe} - C_{pi})C_a$.

The diagrams for the characteristic loads are shown in Figure 9.5 below.

Table 9.1 Wind loads

Element	Eff. Heiglit (m)	Factor S_b	Effective speed V_e	Dynamic pressure q	External pressure coefficient, C_{pe} $\alpha = 0°$		C_{pe} $\alpha = 90°$ Side
					Windward	Leeward	
Roof	10	1.58	38.2	0.89	−0.25	−0.45	−0.6
Walls	6	1.40	33.9	0.70	+0.65	−0.15	−0.8

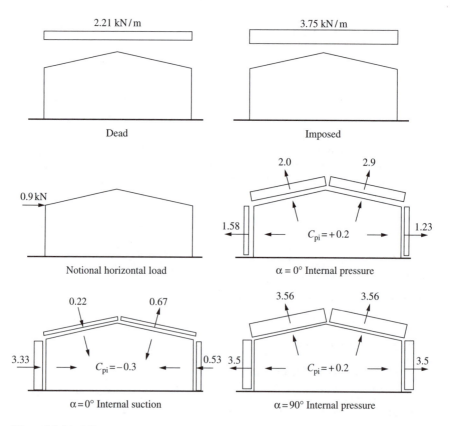

Figure 9.5 Load diagrams

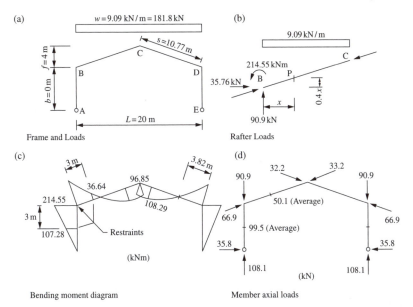

(a) Frame and Loads

(b) Rafter Loads

(c) Bending moment diagram

(d) Member axial loads

Figure 9.6 Forces and moment diagrams

(3) Analysis

The manual analysis for the design dead + imposed load case is set out here. Bending moment diagrams are given for the separate load cases in Figure 9.6.

Frame constants (Refer to Figure 9.6(a))

$$
\begin{aligned}
K &= h/s & &= 0.557, \\
\phi &= f/h & &= 0.667, \\
m &= 1 + \phi & &= 1.667, \\
B &= 2(K + 1) + m & &= 4.781, \\
C &= 1 + 2m & &= 4.334 \\
N &= B + mC = & &= 12.006.
\end{aligned}
$$

Moments and reactions

$$
\begin{aligned}
M_B &= wL^2(3 + 5m)/16N & &= -214.55 \, \text{kN m}, \\
M_C &= (wL^2/8) + mM_B & &= 96.85 \, \text{kN m}, \\
H &= 214.55/6 & &= 35.76 \, \text{kN}
\end{aligned}
$$

Referring to Figure 9.6(b), the moment at any point P in the rafter is

$$ M_p = 90.9x - 214.55 - 35.76 \times 0.4x - 9.09x^2/2 $$

Put $M_p = 0$ and solve to give $x = 3.55 \, \text{m}$,
Put $dM_p/dx = 0$ and solve to give $x = 8.43 \, \text{m}$,
and maximum sagging moment $= 108.29 \, \text{kN m}$.

Thrust at B:

$$ = 90.9 \times 4/10.77 + 35.76 \times 10/10.77 = 66.9 \, \text{kN}. $$

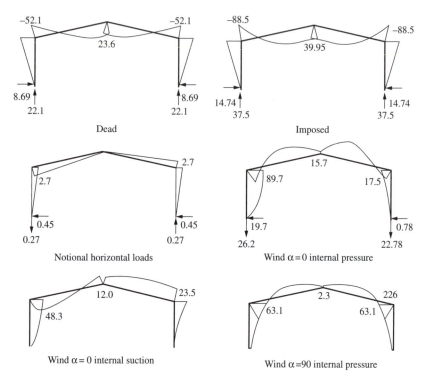

Figure 9.7 Bending moments (kN m) and reactions (kN)

The bending moment diagram and member axial loads are shown in Figure 9.6(c) and (d), respectively. Other values required for design are given in the appropriate figures. The maximum design moments for the separate load cases shown in Figure 9.5 and moment diagram in Figure 9.7 are given for comparison:

Maximum negative moment at the eaves:

(1) $1.0 \times$ dead $+ 1.6 \times$ imposed

$$M_B = -[1.4 \times 52.1 + 1.6 \times 88.5]$$
$$= -214.54 \text{ kN m.}$$

(2) 1.2 [dead $+$ imposed $+$ wind internal suction]

$$M_D = -1.2 [52.1 + 88.5 + 23.5]$$
$$= -196.92 \text{ kN m.}$$

(3) $1.4 \times$ dead $+ 1.6 \times$ imposed $+$ notional horizontal load

$$M_D = -[1.4 \times 52.1 + 1.6 \times 88.5 + 2.7] = 217.24 \text{ kN m.}$$

Maximum reversed moment at the eaves:

$$- 1.0 \times \text{dead} - 1.4 \times \text{wind internal pressure}$$
$$M_B = -52.16 + 1.4 \times 63.1 = 36.2 \text{ kN m.}$$

(4) Column design

Trial section

Try 457×152 UB 60, Grade 275, uniform throughout:

$A = 75.9\,\text{cm}^2$, $r_x = 18.3\,\text{cm}$, $r_y = 3.23\,\text{cm}$, $S_x = 1280\,\text{cm}^3$,
$u = 0.869$, $x = 37.5$.

Compression resistance – Applied load, $F = 90.9\,\text{kN}$.
In-plane slenderness

Fraser's formula

$$L_E/L = 2 + 0.45(2 \times 10.77/6) = 3.62.$$

Using Figure E2, BS 5950-1: 2000:

$$k_1 = \frac{1/6}{1/6 + 1/21.54} = 0.782.$$
$$k_2 = 1.0$$
$$L_E/L = 3.5.$$

Slenderness $L_{Ex}/r_x = 3.62 \times 6000/183 = 118.7$.
Out-of-plane slenderness with restraint at the mid-height of the column as shown in Figure 9.6(c):

Slenderness	$L_{Ey}/r_y = 3000/32.3 = 92.9 \quad < L_{Ex}/r_x$,
Compressive strength	$p_{cx} = 121\,\text{N/mm}^2$ – Table 24(a), BS 5950-1: 2000,
	$p_{cy} = 154\,\text{N/mm}^2$ – Table 24(b), BS 5950-1: 2000.
Compressive resistance	$P_{cx} = 121 \times 75.9/10 = 918.4\,\text{kN}$.
	$P_{cy} = 154 \times 75.9/10 = 1168.9\,\text{kN}$.
Moment capacity	$M_c = 275 \times 1280 \times 10^{-3} = 352\,\text{kN m}$.

Buckling resistance moment:
Applied moment $M = 217.24\,\text{kN}$.

The bending moment diagram is shown in Figure 9.6(c).

Equivalent moment factor, $m_{LT} = 0.8$ for $\beta = 0.5$—Table 18 of the code.

Slenderness factor for $\lambda/x = 92.9/37.5 = 2.48$
$v = 0.931$—Table 19 of the code.

Equivalent slenderness

$$\lambda_{LT} = 0.869 \times 0.931 \times 92.9 \times 1.0 = 75.2.$$

Bending strength:
$p_b = 175.6\,\text{N/mm}^2$—Table 16 of the code.
Buckling resistance moment:
$M_b = 175.6 \times 1280 \times 10^{-3} = 224.8\,\text{kN m}$.

Interaction expressions

Cross-section capacity

$$\frac{90.9}{275 \times 75.9 \times 10^{-1}} + \frac{217.24}{352} = 0.66$$

Member buckling capacity

$$\frac{90.9}{918.4} + \frac{0.6 \times 217.24}{275 \times 1120 \times 10^{-3}} = 0.52$$

$$\frac{90.9}{1168.9} + \frac{0.8 \times 217.24}{224.8} = 0.85$$

The section is satisfactory.

(5) Rafter design check

Compression resistance, $F = 66.9\,\text{kN}$.
The average compressive forces for the rafter and column are shown in Figure 9.6(d).
In-plane slenderness

$$\frac{L_E}{L_{R/2}} = \frac{3.62 \times 2 \times 6}{21.54} \sqrt{\frac{99.5}{50.1}} = 2.84.$$
$$L_{Ex}/r_x = 2.84 \times 10770/183 = 167.$$

Out-plane slenderness: $L_{Ey} = 3000\,\text{mm}$—Figure 9.6(c)

$L_{Ey}/r_y = 3000/32.3 = 92.9.$

$p_{cx} = 65.7\,\text{N/mm}^2$—Table 24(a) of the code,

$p_{cy} = 154\,\text{N/mm}^2$—Table 24(b) of the code,

$P_{cx} = 65.7 \times 75.9/10 = 498.7\,\text{kN},$

$P_{cy} = 154 \times 75.9/10 = 1168.9\,\text{kN}.$

Buckling resistance moment
The bending moment diagram is shown in Figure 9.6(c).
Equivalent moment factor, $m_{LT} = 0.67$ for $\beta = 0.17$—Table 18 of the code.
Equivalent moment factor for flexural buckling, $m = 0.37$—Table 18 of the code.
Slenderness factor for $\lambda/x = 92.9/37.5 = 2.48.$
$v = 0.931$—Table 19 of the code.

Equivalent slenderness

$$\lambda_{LT} = 0.869 \times 0.931 \times 92.9 \times 1.0 = 75.2.$$

Bending strength:

$$p_b = 175.6 \, \text{N/mm}^2\text{—Table 16 of the code.}$$

Buckling resistance moment:

$$M_b = 175.6 \times 1280 \times 10^{-3} = 224.8 \, \text{kN m.}$$

Interaction expressions

Cross-section capacity

$$\frac{66.9}{275 \times 75.9 \times 10^{-1}} + \frac{217.24}{352} = 0.65$$

Member buckling capacity

$$\frac{66.9}{498.7} + \frac{0.37 \times 217.24}{275 \times 1120 \times 10^{-3}} = 0.40$$

$$\frac{66.9}{1168.9} + \frac{0.67 \times 217.24}{224.8} = 0.70.$$

The section is satisfactory.

9.3 Plastic design

9.3.1 Code provisions

BS 5950 states in Section 5.2.3 that plastic design may be used in the design of structures and elements provided that the following main conditions are met:

(1) The loading is predominantly static.
(2) Structural steels with stress–strain diagrams as shown in Figure 3.1 are used. The plastic plateau permits hinge formation and rotation necessary for moment redistribution at plastic moments to occur.
(3) Member sections are plastic where hinges occur. Members not containing hinges are to be compact.
(4) Torsional restraints are required at hinges and within specified distances from the hinges.

Provisions regarding plastic design of portals are given in the code in Section 5.5.3. Other important provisions deal with overall sway stability and column and rafter stability will be discussed later.

9.3.2 Plastic analysis—uniform frame members

Plastic analysis is set out in books such as that by Horne and only plastic analysis for the pinned-base portal is discussed here. The plastic hinge (the formation of which is shown in Figure 4.8) is the central concept. This rotates to redistribute moments from the elastic to the plastic moment distribution.

Referring to Figure 9.8(a), as the load is increased hinges form first at the points of maximum elastic moments at the eaves. Rotation occurs at the eaves' hinges with thereafter acting like a simply supported beam, taking more load until two further hinges form near the ridge, when the rafter collapses. The plastic bending moment diagram and collapse mechanism are shown in Figure 9.8. In general, the number of hinges required to convert the portal to a mechanism is one more than the statical indeterminacy. With unsymmetrical loads such as dead b wind load, two hinges only form to cause collapse.

For the location of the hinges to be correct, the plastic moment at the hinges must not be exceeded at any point in the structure. That is why, in Figure 9.8(b), two plastic hinges form at each side of the ridge and not one only at the ridge at collapse. The critical mechanism is the one which gives the lowest value for the collapse load. The collapse mechanism which occurs depends on the form of loading.

Plastic analysis for the pinned-base portal is carried out in the following stages:

(1) The frame is released to a statically determinate state by inserting rollers at one support.
(2) The free bending moment diagram is drawn.
(3) The reactant bending moment diagram due to the redundant horizontal reaction is drawn.
(4) The free and reactant moment diagrams are combined to give the plastic bending moment diagram with sufficient hinges to cause the frame or part

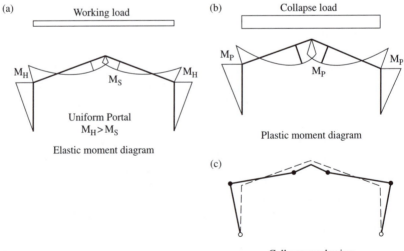

Figure 9.8 Collapse mechanism of pinned base portal frame

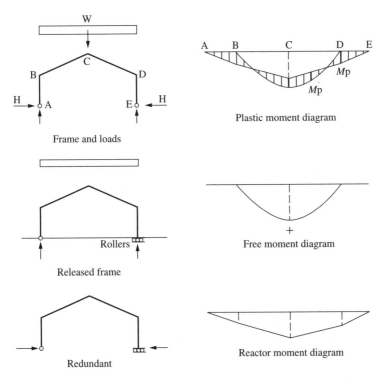

Frame and loads

Plastic moment diagram

Released frame

Free moment diagram

Redundant

Reactor moment diagram

Figure 9.9 Plastic moment diagram

of it (e.g. the rafter) to collapse. As mentioned above, the plastic moment must not be exceeded.

The process of plastic analysis for the pinned-base portal is shown for the case of dead and imposed load on the roof in Figure 9.9. The frame is taken to be uniform throughout and the bending moment diagrams are drawn on the opened-out frame. The case of dead + imposed + wind load is treated in the design example.

The exact location of the hinge near the ridge must be found by successive trials or mathematically if the loading is taken to be uniformly distributed. Referring to Figure 9.10, for a uniform frame, hinge X is located by

$$g = h + 2fx/L$$

The free moment at X in the released frame is

$$M_x = wLx/2 - wx^2/2$$

The plastic moments at B and X are equal, i.e.

$$M_p = Hh = M_x - Hg$$

Put $dH/dx = 0$ and solve for x and calculate H and M_p. Symbols used are shown in Figure 9.10.

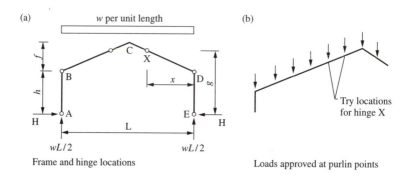

Figure 9.10 Plastic hinge locations

If the load is taken to be applied at the purlin points as shown in Figure 9.10(b) the hinge will occur at a purlin location. The purlins may be checked in turn to see which location gives the maximum value of the plastic moment M_p.

9.3.3 Plastic analysis—haunches and non-uniform frame

Haunches are provided at the eaves and ridge primarily to give a sufficient lever arm to form the bolted joints. The haunch at the eaves causes the hinge to form in the column at the bottom of the haunch. This reduces the value of the plastic moment when compared with the analyses for a hinge at the eaves intersection. The haunch at the ridge will not affect the analyses because the hinge on the rafter forms away from the ridge. The haunch at the eaves is cut from the same UB as the rafter. The depth is about twice the rafter depths and the length is often made equal to span/10.

It is also more economical to use a lighter section for the rafter than for the column. The non-uniform frame can be readily analysed as discussed below. It is also essential to ensure that the haunched section of the rafter at the eaves remains elastic. That is, the maximum stress at the end of the haunch must not exceed the design strength, p_y:

$$p_y = \frac{F}{A} + \frac{M}{Z}$$

where

F is the axial force,
M the moment,
A the area, and
Z the elastic modulus.

The analysis of a frame with haunched rafter and lighter rafter than column section is demonstrated with reference to Figure 9.11(a).

Frame dimensions are shown in Figure 9.11.

Hinge in column: $M_p = He$

Hinge in rafter: $q M_p = M_x - Hg = qHe$

where

M_p is the plastic moment of resistance of the column,
$q M_p = $ plastic moment of resistance of the rafter,

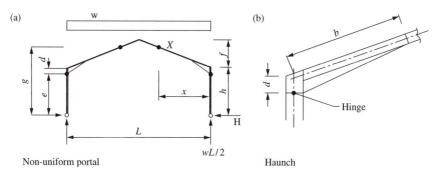

(a)   w

X

d

g

e

f

h

x

L

wL/2

Non-uniform portal

(b)   b

d

Hinge

H

Haunch

Figure 9.11 Non-uniform portal frame

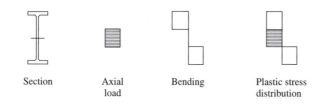

Section   Axial   Bending   Plastic stress
           load   distribution

Figure 9.12 Stress diagram

q = normally 0.6–0.7,
M_x = free moment at X in the released frame and $g = h + 2fx/L$.

Put $dH/dx = 0$ and solve to give x and so obtain H and M_p.

9.3.4 Section design

At hinge locations, design is made for axial load and plastic moment. The following two design procedures can be used:

(1) Simplified method

Local capacity check:

$$\frac{F}{A_g p_y} + \frac{M}{M_c} \leq 1$$

where

F and M are applied load and plastic moment, respectively,
M_c is the moment capacity and
A_g the gross area.

(2) Exact method

Axial load reduces the plastic moment of resistance of a section. The bending moment is resisted by two equal areas extending inwards from the edges. The central area resists axial load and this area may be confined to the web or extend into the flange under heavy load. The stress diagrams are shown in Figure 9.12.

The formulae for calculating the reduced plastic moduli of sections subjected to axial load and moment are given in *Steelwork Design to BS 5950, Volume I* and the calculation procedure is as follows:

$$n = F/A_g p_y$$

where

F is the applied axial load
A_g the gross area.

Change values of n are given for each section. For lower values of n, the neutral axis lies in the web and the reduced plastic modulus:

$$S_r = k_1 - k_2 n^2$$

where values of k_1 and k_2 are given in the *Steelwork Design to BS 5950, Volume I*. A formula is also given for upper values of n. Reduced moment capacity, $M_r = S_r p_y > M$, applied moment.

9.4 In-plane stability

The in-plane stability of a portal frame should be checked under each load combination. Section 5.5.4, BS 5950-1: 2000 gives three procedures for checking the overall sway stability of a portal:

(a) Sway-check method given in Section 5.5.4.2. in the code.
(b) Amplified moments method given in Section 5.5.4.4. in the code.
(c) Second-order analysis given in Section 5.5.4.5. in the code.

These ensure that the elastic buckling load is not reached and that the effects of additional moments due to deflection of the portal are taken into account.

9.4.1 Sway-check method

The sway-check method may be used to verify the in-plane stability of portal frames in which each bay satisfies the following conditions:

(a) span, L does not exceed five times the mean height h of the columns;
(b) height, h_r of the apex above the tops of the columns does not exceed 0.25 times the span, L.

(1) Limiting horizontal deflection at eaves

For gravity loads (load combination 1), the horizontal deflection calculated by linear elastic analysis at the top of the columns due to the notional horizontal loads without any allowance for the stiffening effects of cladding should not exceed $h/1000$, where h is the column height. The notional loads are 0.5% of the factored roof dead and imposed loads applied at the column tops.

(2) Limiting span/depth ratio of the rafter

The $h/1000$ sway criterion for gravity loads may be assumed to be satisfied if:

$$\frac{L_b}{D} \le \frac{44L}{\Omega h} \left(\frac{\rho}{4 + \rho L_r / L} \right) \left(\frac{275}{p_{yr}} \right)$$

in which

$$L_b = L - \left(\frac{2D_h}{D_s + D_h} \right) L_h$$

For single bay: $\rho = \left(\dfrac{2I_c}{I_r} \right) \left(\dfrac{L}{h} \right)$

For multi-bay: $\rho = \left(\dfrac{I_c}{I_r} \right) \left(\dfrac{L}{h} \right)$

and Ω is the arching ratio, given by:

$$\Omega = \frac{W_r}{W_o}$$

where

D is the cross-section depth of the rafter,
D_h the additional depth of the haunch,
D_s the depth of the rafter, allowing for its slope,
h the mean column height;
I_c the in-plane second moment of area of the column (taken as zero if the column is not rigidly connected to the rafter, or if the rafter is supported on a valley beam),
I_r the in-plane second moment of area of the rafter,
L the span of the bay,
L_b is the effective span of the bay.
L_h the length of a haunch,
L_r the total developed length of the rafters,
p_{yr} the design strength of the rafters in N/mm^2;
W_o the value of W_r for plastic failure of the rafters as a fixed-ended beam of span L, and
W_r the total factored vertical load on the rafters of the bay.

(3) Horizontal load

For load combinations that include wind loads or other significant horizontal loads, allowance may be made for the stiffening effects of cladding in calculating the notional horizontal deflections. Provided that the $h/1000$ sway criterion is satisfied for gravity loads, then for load cases involving horizontal loads, the required load factor λ_r for frame stability should be determined using:

$$\lambda_r = \frac{\lambda_{sc}}{\lambda_{sc} - 1}$$

in which λ_{sc} is the smallest value, considering every column, determined from:

$$\lambda_{sc} = \frac{h_i}{200\delta_i}$$

using the notional horizontal deflections δ_i for the relevant load case.

If $\lambda_{sc} < 5$, second order analysis should be used.

λ_{sc} may be approximate using:

$$\lambda_{sc} = \frac{200DL}{\Omega h L_b} \left(\frac{\rho}{4 + \rho L_r/L}\right) \left(\frac{275}{p_{yr}}\right).$$

9.4.2 Amplified moments method

For each load case, the in-plane stability of a portal frame may be checked using the lowest elastic critical load factor λ_{cr} for that load case. This should be determined taking account of the effects of all the members on the in-plane elastic stability of the frame as a whole.

In this method, the required load factor λ_r for frame stability should be determined from the following:

$$\lambda_{cr} \geq 10: \quad \lambda_r = 1.0$$

$$10 \geq \lambda_{cr} \geq 4.6: \quad \lambda_r = \frac{0.9\lambda_{cr}}{\lambda_{cr} - 1}$$

If $\lambda_{cr} < 4.6$, the amplified moments method should not be used.

For pinned base portal:

$$\lambda{cr} = \frac{3EI_r}{S\left[(1 + (1.2/R))\, P_c h + 0.3 P_r S\right]}$$

For fixed base portal:

$$\lambda_{cr} = \frac{5E(10 + R)}{5 P_r S^2/I_r + 2R P_c h^2/I_c}$$

where

$$R = \frac{I_c L_r}{I_r h},$$

P_c is the axial compression in column from elastic analysis, P_r the axial compression in rafter from elastic analysis and S the rafter length along the slope (eaves to apex).

9.4.3 Second-order analysis

The in-plane stability of a portal frame may be checked using either elastic or elastic–plastic second order analysis. When these methods are used, the required load factor λ_r for frame stability should be taken as 1.0. Further information on second-order analysis can be found on *In-plane stability of portal frames to BS 5950–1: 2000 by SCI.*

9.5 Restraints and member stability

9.5.1 Restraints

Restraints are required to ensure that:

(1) Plastic hinges can form in, the deep 1-sections used.
(2) Overall flexural buckling of the column and rafter about the minor axis does not occur.
(3) There is no lateral torsional buckling of an unrestrained compression flange on the inside of members.

The code requirements regarding restraints and member satiability are set out below. A restraint should be capable of resisting 2.5 per cent of the compressive force in the member or part being restrained.

9.5.2 Column stability

The column contains a plastic hinge near the top at the bottom of the haunch. Below the hinge, it is subjected to axial load and moment with the inside flange in compression-. The code states in Section 5.3.2 that torsional restraints (i.e. restraints to both flanges) must be provided at or within member depth $D/2$ from a plastic hinge.

In a member containing a plastic hinge the maximum distance from the restraint at the hinge to the adjacent restraint depends on whether or not restraint to the tension flange is taken into account. The following procedures apply:

(1) Restraint to tension flange not taken into account

This is the conservative method where the distance from the hinge restraint to the next restraint is given by:

$$L_m = \frac{38r_y}{\left[f_c/130 + (x/36)^2 \left(p_y/275\right)^2 \right]^{0.5}}$$

where f_c is the compressive stress (in N/mm^2) due to axial force, p_y the design strength (in N/mm^2), r_y the radius of gyration about the minor axis and x the torsional index.

When this method is used, no further checks are required. The locations of restraints at hinge H and at G, L, below H are shown in Figure 9.13(a). It may be necessary to introduce a further restraint at F below G, in which case column lengths GF and FA would be checked for buckling resistance for axial load and moment. The effective length for the x–x axis may be estimated for the portal as set out in Section 9.2.3. The *Steelwork Design Guide to BS 5950*

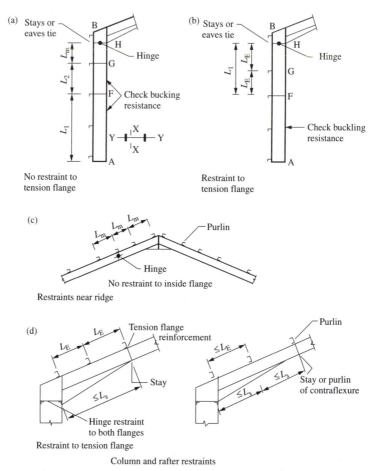

Figure 9.13 Column and rafter restraints

takes the effective length for the $x-x$ axis as HA in Figure 9.13, the distance between the plastic hinge and base. The effective lengths for the $y-y$ axis are GF and FA. Note that compliance with the sway stability check ensures that the portal can safely resist in-plane buckling and additional moments due to frame deflections.

(2) Restraint to tension flange taken into account

A method for determining spacing of lateral restraints taking account of restraint to the tension flange is given in Clause 5.3.4 of BS 5950-1: 2000. A more rigorous method is given in Annex G. The formula from Clause 5.3.4 gives the following limit:

for steel grade S275:

$$L_s = \frac{620r_y}{K_1 \left[72 - (100/x)^2\right]^{0.5}}$$

for steel grade S355:

$$L_s = \frac{645 r_y}{K_1 \left[94 - (100/x)^2 \right]^{0.5}}$$

—for an un-haunched segment: $K_1 = 1.00$;
—for a haunch with $D_h/D_s = 1$: $K_1 = 1.25$;
—for a haunch with $D_h/D_s = 2$: $K_1 = 1.40$;
—for a haunch generally: $K_1 = 1 + 0.25 (D_h/D_s)^{2/3}$.

The code specifies that the buckling resistance moment M_b calculated using an effective length L_E equal to the spacing of the tension flange restraints must exceed the equivalent uniform moment for that column length. The restraints are shown in the Figure 9.13(b). The column length AF is checked as set out above.

9.5.3 Rafter stability near ridge

The tension flange at the hinge in the rafter near the ridge is on the inside and no restraints are provided. In the portal, two hinges form last near the ridge. A purlin is required at or near the hinge and purlins should be placed at a distance not exceeding L_m on each side of the hinge. The purlin arrangement is shown in Figure 9.13(c). In wide-span portals (say, 30 m or over), restraints should be provided to the inside flange.

9.5.4 Rafter stability at haunch

The requirements for rafter stability are set out in Clause 5.3.4 of the code. The haunch and rafter are in compression on the inside from the eaves for a distance to the second or third purlin up from the eaves where the point of contraflexure is usually located:

(1) Restraint to tension flange not taken into account

The maximum distance between restraints to the compression flange must not exceed L_m as set out in column stability above or that to satisfy the overall buckling expression given in Section 4.8.3.3.1 of the code. For the tapered member the minimum value of r_y and maximum value of x are used. No restraint is to be assumed at the point of contraflexure.

(2) Restraint to tension flange taken into account

When the tension flange is restrained at intervals by the purlins, possible restraint locations for the compression flange are shown in Figure 9.13(d). The code requirements are:

(a) The distance between restraints to the tension flange, L_E must not exceed L_m or, alternatively, the interaction expression for overall buckling given in Section 4.8.3.3.1 of the code based on an effective length L_E must be satisfied.

(b) The distance between restraints to the compression flange must not exceed L_s specified in Section 5.3.4 in the BS 5950-1: 2000.

The following provisions are set out in the code:

(1) The rafter must be a universal beam.
(2) The depth of haunch must not exceed three times the rafter depth.
(3) The haunch flange must not be smaller than the rafter flange.

If these conditions are not met, the rigorous method given in Annex G of the code should be used.

The code also specifies that for the purlins to provide restraint to the top flange, they must be connected by two bolts and have a depth not less than one quarter of the rafter depth. A vertical lateral restraint should not be automatically assumed at the point of contraflexure without provision of stays.

9.6 Serviceability check for eaves deflection

BS 5950 specifies in Table 8 that the deflection at the column top in a single-storey building is not to exceed height/300 unless such deflection does not damage the cladding. The deflections to be considered are due to the unfactored imposed and wind loads. If necessary, an allowance can be made for dead load deflections in the fabrication.

A formula for horizontal deflection at the eaves due to uniform vertical load on the roof is derived. Deflections for wind load should be taken from a computer analysis.

Referring to Figure 9.14, because of symmetry of the frame and loading the slope at the ridge does not change. The slope θ at the base is equal to the area of the M/EI diagram between the ridge and the base given by

$$\theta = \frac{1}{E I_R} \left[\frac{w L^2 s}{24} + Hs \left(h + \frac{f}{2} \right) - \frac{VLs}{4} \right] + \frac{Hh^2}{2EI_c}$$

The deflection at the eaves is then

$$\delta_B = h\theta - \frac{Hh^3}{6EI_c}$$

where

H is the horizontal reaction at base and
V the vertical reaction at base.
w represents characteristic imposed load on the roof and
I_c and I_R are second moments of area of the column and rafter, respectively.

Frame dimensions are shown in Figure 9.14.

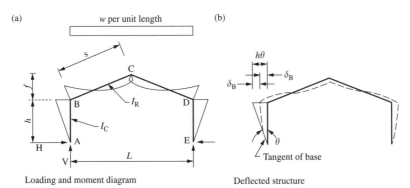

(a)

w per unit length

(b)

Loading and moment diagram

Deflected structure

Figure 9.14 Deflection for portal frame

9.7 Design of joints

9.7.1 Eaves joint

The eaves joint arrangement is shown in Figure 9.15(a). The steps in the joint design check are:

(1) Joint forces

Take moment about X:

$$\frac{Vd}{2} - M_p + Ta = 0$$

Bolt tension:

$$T = \left(M_p - \frac{Vd}{2}\right) \bigg/ a$$

Compression:

$$C = T + H$$

Haunch flange force:

$$F = C \sec \phi$$

(2) Bolt design

Tension:

$$F_T = \frac{T}{4},$$

Shear:

$$F_S = \frac{V}{8}.$$

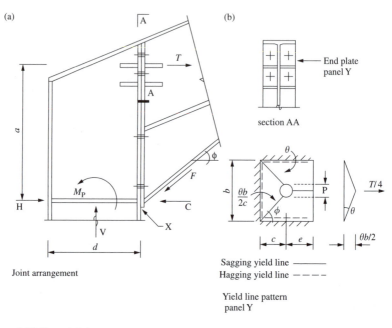

(a)

(b)

Joint arrangement

Sagging yield line ———
Hagging yield line — — — —

Yield line pattern
panel Y

Figure 9.15 Eaves joint

for a given bolt size with capacities P_T in tension and P_S in shear. The interaction expression is

$$\frac{F_S}{P_S} + \frac{F_T}{P_T} \leq 1.4.$$

(3) Column flange check and end plate design

Adopt a yield line analysis (see Horne and Morris in the further reading at the end of this chapter). The yield line pattern is shown in Figure 9.15(b) for one panel of the end plate or column flange. The hole diameter is v.

Work done by the load

$$= \frac{T}{4} \frac{b\theta}{2}$$

Work done in the yield lines

$$4m\theta \left(c + e \right) - mv\theta \left(1 + \cos \phi \right) + \frac{mb}{c} \left[b - \frac{v}{2} \sin \phi \right]$$

Equate the expressions and solve for m. The plate thickness required

$$t = \left(\frac{4m}{p_y} \right)^{0.5}$$

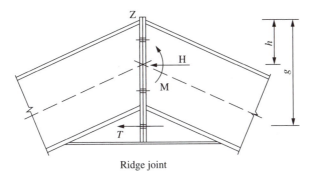

Ridge joint

Figure 9.16 Ridge joint

(4) Haunch flange

Flange force $F < p_y \times$ flange area. The haunch section is checked for axial load and moment in the haunch stability check.

(5) Column stiffener

Design the stiffener for force C. See Section 5.3.7 of the book.

(6) Column web

Check for shear V. See Section 4.6.2.

(7) Column and rafter webs

These webs are checked for tension T. The small stiffeners distribute the load.

(8) Welds

The fillet welds from the end plate to rafter and on the various stiffeners must be designed.

The main check calculations are shown in the example in Section 9.8.

9.7.2 Ridge joint

The ridge joint is shown in Figure 9.16. The bolt forces can be found by taking moments about Z:

$$T = \frac{(M - Hh)}{g}$$

where H is the horizontal reaction in the portal and M the ridge moment.

Joint dimensions are shown in the figure. Other checks such as for end plate thickness and weld sizes are made in the same way as for the eaves joint.

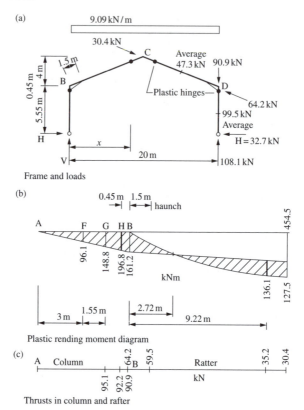

(a)

Frame and loads

(b)

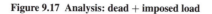

Plastic rending moment diagram

(c)

Thrusts in column and rafter

Figure 9.17 Analysis: dead + imposed load

9.8 Design example of a portal frame

9.8.1 Specification

Redesign the pinned-base portal specified in Section 9.2.6 using plastic design. The portal shown in Figure 9.17(a) has haunches at the eaves 1.5 m long and 0.45 m deep. The rafter moment capacity is to be approximately 75% of that of the column.

9.8.2 Analysis—dead and imposed load

The frame dimensions, loading and plastic hinge locations are shown in Figure 9.17(a). The plastic moments in the column at the bottom of the haunch and in the rafter at x from the eaves are given by the following expressions:

Column: $M_p = 5.55H$.

Rafter: $0.75M_p = 90.9x - 9.09x^2/2 - H(6 + 0.4x)$.

Reduce to give

$$H = \frac{90.9x - 4.55x^2}{10.16 + 0.4x}$$

Put $dh/dx = 0$ and collect terms to give

$$x^2 + 50.7x - 507.4 = 0$$

Solve $x = 8.56$ m,
   $H = 32.7$ kN,
   $M_p = 181.5$ kN m—column,
$0.75 M_p = 136.1$ kN m—rafter.

The plastic bending moment diagram with moments needed for design is shown in Figure 9.17(b) and the co-existent thrusts in (c).

9.8.3 Analysis—dead + imposed + wind load

The frame is analysed for the load case:

1.2 [dead + imposed + wind internal suction]

The roof wind loads acting normally are resolved vertically and horizontally- and added to the dead and imposed load. The frame and hinge locations are shown in Figure 9.18(a) and the released frame, loads and reactions in (b).

The free moment in the rafter at distance x from the eaves point can be expressed by

$$M_x = 66.5x - 57.24 - 3.52x^2.$$

The equations for the plastic moments at the hinges are:
Columns: $M_p = 5.55H + 2.59$.
Rafter: $0.75 M_p = 66.5x - 57.24 - 3.52x^2 - H(6 + 0.4x)$

This gives

$$H = \frac{66.5x - 59.18 - 3.52x^2}{10.16 + 0.4x}$$

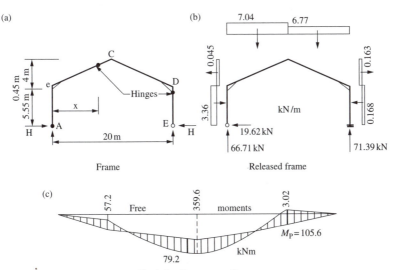

Figure 9.18 Analysis: Dead + imposed + windload

Put $dh/dx = 0$ and solve to give $x = 8.39$ m.

$$H = 18.6 \text{ kN},$$
$$M_p = 105.1 \text{ kN}.$$

The plastic bending moment diagram is shown in Figure 9.18(c). The moments are less than for the dead + imposed load case.

9.8.4 Column design

The design actions at the column hinge are

$$M_p = 181.8 \text{ kN m},$$
$$F = 92.2 \text{ kN},$$
$$S_x = 181.8 \times 10^3/275 = 661.1 \text{ cm}^3.$$

Try 406 × 140 UB 46

$$S_x = 888 \text{ cm}^3, \quad A = 59 \text{ cm}^2, \quad r_x = 16.3 \text{ cm}, \quad r_y = 3.02 \text{ cm},$$
$$u = 0.87, x = 38.8, \quad I_x = 15\,600 \text{ cm}^4.$$
$$n = 92.2 \times 10/(59 \times 275) = 0.056 < 0.444.$$

Reduced plastic modulus:

$$S_{rx} = 888 - 1260n^2 = 883.9 \text{ cm}^3.$$

The section is satisfactory.

9.8.5 Rafter section

The design actions at the rafter hinge are:

$$M_p = 136.1 \text{ kN m}$$
$$F = 35.2 \text{ kN},$$
$$S = 136.1 \times 10^3/275 = 494.9 \text{ cm}^3.$$

Try 356 × 127 UB 39

$$S_x = 654 \text{ cm}^3, \quad A = 49.4 \text{ cm}^2, \quad r_x = 14.3 \text{ cm}, \quad r_y = 2.69 \text{ cm},$$
$$Z_x = 572 \text{ cm}^3, \quad I_x = 10\,100 \text{ cm}^4, \quad x = 34.3.$$
$$n = 35.5 \times 10/(49.4 \times 275) = 0.026 < 0.437,$$
$$S_{rx} = 654\,0.43 = 653.6 \text{ cm}^3.$$

The section is satisfactory.

Check that the rafter section at the end of the haunch remains elastic under factored loads. The actions are

$$M = 96.9 \text{ kN m},$$
$$F = 59.5 \text{ kN}$$

$$\text{Maximum stress} = \frac{59.5 \times 10}{49.4} + \frac{96.9 \times 10^3}{572} = 181.4 \text{ N/mm}^2.$$

This is less than $p_y = 275$ N/mm^2. The section remains elastic.

9.8.6 Column restraints and stability

A stray is provided to restrain the hinge section at the eaves. The distance to the adjacent restraint using the conservative method is:

$$L_m = \frac{38 \times 30.2}{\left[\frac{92.2 \times 10}{59 \times 130} + \left(\frac{38.8}{36}\right)^2 \left(\frac{275}{275}\right)^2\right]^{0.5}} = 1013.7 \text{ mm}.$$

The arrangement for the column restraints is shown in Figure 9.19(a). The column is checked between the second and third restraints G and F over a length of 1.55 m. The moments and thrusts are shown in Figure 9.17.

$$M_F = 98.1 \text{ kN m},$$
$$M_G = 148.8 \text{ kN m}$$
$$F_G = 95.1 \text{ kN}.$$

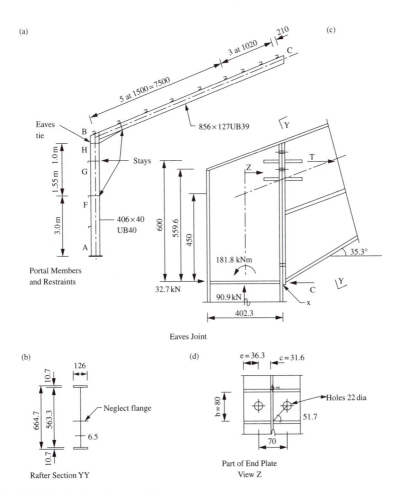

Figure 9.19 Restraints and eaves joint

The effective lengths and slenderness ratios are

$$k_c = \left[2 + \frac{0.45 \times 21.54 \times 15600}{6 \times 10100}\right] = 4.49$$

$L_{EX}/r_x = 4.49 \times 6000/163 = 165.3$.
Note that the *Steelwork Design Guide to BS 5950* would use $L_{EX}/r_x = 5550/163 = 34$ in this check

$L_{Ey}/r_y = 1550/30.2 = 51.3$,
$p_{cx} = 66.7\,\text{N/mm}^2$—Table 24(a) in the code,
$p_{cy} = 235.5\,\text{N/mm}^2$—Table 24(b) in the code,
$P_{cx} = 66.7 \times 59/10 = 393.5\,\text{kN}$,
$P_{cx} = 235.5 \times 59/10 = 1389.5\,\text{kN}$,
$\beta = 0, m = 0.6$—Table 26 in the code, and
$\beta = 98.1/148.8 = 0.66, m_{LT} = 0.86$—Table 18 in the code,
$\lambda/x = 56.3/38.8 = 1.45, v = 0.972$—Table 19 in the code,
$\lambda_{LT} = 0.87 \times 0.972 \times 56.3 \times 1.0 = 47.6$,
$p_b = 243.8\,\text{N/mm}^2$—Table 16 in the code, and
$M_b = 243.8 \times 888 \times 10^{-3} = 216.5\,\text{kN m}$.

Member buckling capacity:

$$\frac{95.1}{393.5} + \frac{0.6 \times 196.2}{275 \times 778 \times 10^{-3}} = 0.79$$

$$\frac{95.1}{1389.5} + \frac{0.86 \times 148.8}{216.5} = 0.66$$

Satisfactory.

Note that the in-plane stability or sway stability is checked by the code procedure in (8) below.

9.8.7 Rafter restraints and stability

The rafter section at the caves is shown in Figure 9.19(b). The centre flange is neglected and the section properties are calculated to give

$A = 64.9\,\text{cm}^2$,

$S_x = 1353\,\text{cm}^3$,

$r_x = 2.35\,\text{cm}$.

$$x = h_s \left[\frac{B_c T_c + B_t T_t + dt}{B_c T_c^3 + B_t T_t^3 + dt^3}\right]^{0.5} \text{—Annex B2.4.1 in the BS 5590-1: 2000.}$$

$$x = 59.4 \left[\frac{12.6 \times 1.07 + 12.6 \times 1.07 + 58.33 \times 0.65}{12.6 \times 1.07^3 + 12.6 \times 1.07^3 + 58.33 \times 0.65^3}\right]^{0.5} = 69.9.$$

The rafter stability is checked taking account of restraint to the tension flange using Clause 5.3.4 in the code.

Depth of haunch/depth of rafter $= 380.6/337 = 1.13$

$$K_1 = 1 + 0.25(1.13)^{2/3} = 1.27$$
$$L_s = \frac{620 \times 23.5}{1.27\left[72 - (100/69.9)^2\right]^{0.5}} = 1372 \text{ mm.}$$

Provide stays at 1372 mm for the haunch.

Check overall buckling in accordance with Clause 4.8.3.3.1 using an effective length L_E of 1372 mm for the haunch:

$$L_{Ey}/r_y = 1372/23.5 = 58.4,$$
$$\lambda/x = 58.4/69.9 = 0.84, v = 0.993\text{—Table 19 in the code,}$$
$$u = 1.0,$$
$$\lambda_{LT} = 1.0 \times 0.993 \times 58.4 \times 1.0 = 58.0.$$

The smallest value of r_y and the largest value of x have been used.

At the large end of the haunch, at the eaves, consider as a welded section:

$F = 64.2$ kN from Figure 9.17,
$M_x = 196.2$ kN m from Figure 9.17,
$p_b = 187$ N/mm^2 from Table 17 of the code,
$p_c = 204$ N/mm^2 from Table 24(c) of the code, and
$M_b = 187 \times 1353 \times 10^{-3} = 253.0$ kN m.

Interaction expression

$$\frac{64.2 \times 10}{213.4 \times 69.4} + \frac{196.2}{253} = 0.82.$$

At the small end of the haunch, consider as a rolled section 356×127 UB 39 for which $A = 49.4$ cm^2, $S_x = 654$ cm^3. The actions at the end of the haunch are:

$F = 59.5$ kN. See section (5) above.
$M_x = 96.9$ kN m,
$p_b = 203.9$ N/mm^2 from Table 16 of the code,
$p_c = 213.4$ N/mm^2 from Table 27(b) of the code and
$M_b = 203.9 \times 654 \times 10^{-3} = 133.4$ kN m.

Interaction expression

$$\frac{59.5 \times 10}{213.4 \times 49.4} + \frac{96.9}{133.4} = 0.79$$

The haunch is satisfactory.

If restraint to the tension flange is not considered:

At the large end of the haunch,

$$L_m = \frac{38 \times 23.5}{\left[\frac{64.2 \times 10}{64.9 \times 130} + \left(\frac{68.6}{36} \right)^2 \left(\frac{275}{275} \right)^2 \right]^{0.5}} = 463.8 \text{ mm.}$$

At the small end of the haunch,

$$L_m = \frac{38 \times 26.9}{\left[\frac{35.2 \times 10}{49.4 \times 130} + \left(\frac{35.3}{36} \right)^2 \left(\frac{275}{275} \right)^2 \right]^{0.5}} = 1030.5 \text{ mm.}$$

An additional purlin must be located at, say, 450 mm from the eaves with stays to the compression flange.

At the hinge location near the ridge the purlins will be spaced at 1020 mm as shown in Figure 9.19(a).

9.8.8 Sway stability

Using sway-check method in section 9.4.1, $L/h < 5$; $h_r < 0.25L$, use limiting span/depth of rafter ratio. The haunch depth 604.7 mm (Figure 9.19(b)) is less than twice the rafter depth, the effective span $L_b = 20$ m:

$$\rho = \left(\frac{2 \times 15600}{10100} \right) \left(\frac{20}{6} \right) = 10.29.$$

$$M_p = 654 \times 275 \times 10^{-3} = 179.9 \text{ kN m—rafter,}$$

$$= W_o L/16.$$

$$W_o = 143.9 \text{ kN.}$$

$$\Omega = 9.09 \times 20/143.9 = 1.26$$

$$\frac{L}{D} = \frac{20000}{352.8} = 56.7 \leq \frac{44 \times 20}{1.26 \times 6} \left(\frac{10.29}{4 + 10.29 \times 21.54/20} \right) \left(\frac{275}{275} \right)$$

$$= 79.4.$$

Satisfactory.

9.8.9 Serviceability—Deflection at eaves

An elastic analysis for the imposed load on the roof of 3.75 kN/m gives $H = 15.26\,\text{kN}$ and $V = 37.5\,\text{kN}$. See Sections 9.2.2 and 9.2.6. From Section 9.6,

$$
\theta = \frac{1}{205 \times 10^3 \times 10100 \times 10^4}
$$

$$
\times \left[\frac{3.75 \times 20000^2 \times 10770}{24} + 15260 \times 10770 \left(600 + \frac{4000}{2} \right) \right]
$$

$$
- \frac{37500 \times 20000 \times 10770}{4}
$$

$$
+ \frac{15260 \times 6000^2}{2 \times 205 \times 10^3 \times 15600 \times 10^4} = 7.068 \times 10^{-3}\,\text{radians}.
$$

Defection at eaves:

$$
S_B = 6000 \times 7.068 \times 10^{-3} - \frac{15260 \times 6000^3}{6 \times 205 \times 10^3 \times 15600 \times 10^4}
$$

$$
= 25.2\,\text{mm} = \text{height}/238.
$$

This exceeds $h/300$ but metal sheeting will accommodate the deflection.

9.8.10 Design of joints

The arrangement for the eaves joint is shown in Figure 9.19(c). Selected check calculations only are given.

(a) Joint forces:

Take moment about X:

$$
T = [181.8 - 90.9 \times 0.402/2]/0.6 = 272.5\,\text{kN}.
$$

(b) Bolts:

$$
F_T = 272/4 = 68.13\,\text{kN}
$$
$$
F_S = 90.0/6 = 15.15\,\text{kN}.
$$

Try 20 mm Grade 8.8 bolt. From the *Steel Design Guide to BS 5950, Volume 1*:

$$
P_T = 110\,\text{kN},
$$
$$
P_S = 91.9\,\text{kN}.
$$

$$
\frac{15.19}{91.9} + \frac{68.13}{110} = 0.78 < 1.4
$$

The bolts are satisfactory.

(c) Column flange:

See Figure 9.19(d) and Section 9.3.8. The yield line analysis gives:

$$\frac{272.5 \times 10^3 \times 80}{8} = 4m \times 67.9 - m \times 22(1 + 0.62)$$

$$+ \frac{m \times 80}{31.6}\left(80 - \frac{22 \times 0.78}{2}\right) = 416.8\,\text{m}$$

$$m = 6537.9\,\text{N mm/mm}.$$

$$t = \left(\frac{4 \times 6537.9}{275}\right)^{0.5} = 9.75\,\text{mm}.$$

The flange thickness 11.2 mm is adequate. The rafter end plate can be made 12 mm thick.

(d) Column web shear:

$$F_v = 0.6 \times 275 \times 402.3 \times 6.9 \times 10^3 = 458.8 kN > 272.8\,\text{kN}$$

Satisfactory.

A similar design procedure is carried out for the ridge joint.

9.8.11 Comments

(a) Wind uplift:

The rafters on portals with low roof angles and light roof dead loads require checking for reverse bending due to wind uplift. Restraints to inside flanges near the ridge are needed to stabilize the rafter. Joints must be also be checked for reverse moments.

The portal frames designed above has a relatively high roof angle and heavy roof dead load. Checks for load, $-1.4 \times \text{wind} + 1.0 \times \text{dead}$, for wind angles $0°$ and $90°$ show that the frame remains in the elastic range and the rafter is stable without adding further restraints. See Figure 9.13.

(b) Eaves joint design:

In designing the joint, the effect of axial load is usually ignored, and the bolts are sized to resist moment only. In the above case, $M = 180.8\,\text{kN m}$ and the bolt tension $T = 75.3\,\text{kN} < 110\,\text{kN}$. The shear force is taken by the bottom two bolts.

9.9 Further reading for portal design

DAVIES, J. M., 'In-plane Stability in Portal Frames', *The Structural Engineer*, **68**, No. 8, April 1990

DAVIES, J. M., 'The Stability of Multi-bay Portal Frames', *The Structural Engineer*, **69**, No. 12, June 1991

FRASER, D. J., 'Effective Lengths in Gable Frames, sway not prevented': *Civil Engineering Transaction, Institution of Engineers, Australia* **CE 22**, No. 3, 1980

FRASER, D. J., 'Stability of Pitched Roof Frames': *Civil Engineering Transaction, Institution of Engineers, Australia* **CE 28**, No. 1, 1986

HORNE, M. R. and MERCHANT, W., *'The stability of frames'*, Oxford, Fergamon Press, 1965

HORNE, M. R., *'Plastic Theory of Structures'*, Nelson, London, 1971

HORNE, M. R. and MORRIS, L. J., *'Plastic Design of Low Rise Frames'*, Collins, London, 1981

In-plane Stability of Portal Frames to BS5950-1: 2000, The Steel Construction Institute, Ascot

Steel Designer's Manual, The Steel Construction Institute, Blackwell Science.

Steelwork Design Guide to BS5950-1: Volume *1, Section Properties and Member Capacities*, The Steel Construction Institute, Ascot 2000.

ɔnnections

0.1 Types of connections

Ɔonnections are needed to join:

a) members together in trusses and lattice girders;
b) plates together to form built-up members;
(c) beams to beams, beams, trusses, bracing, etc. to columns in structural frames, and
(d) columns to foundations.

Some of these typical connections are shown in Figure 10.1. Basic connections are considered in this chapter and end connections for beams and column bases are treated in the chapters on beams and columns, respectively.

Connections may be made by:

- bolting – non-preloaded bolts in standard clearance or oversize holes;
      – preloaded or friction-grip bolt; and
- welding – fillet and butt welds.

10.2 Non-preloaded bolts

10.2.1 Bolts, nuts and washers

The ISO metric 'black' hexagon head ordinary non-preloaded bolt shown in Figure 10.2 with nut and washer is the most commonly used structural fastener in the industry. The bolts, in the three common strength grades given below, are specified in BS 4190: 2001. The specification for ISO metric 'precision' non-preloaded hexagon bolts and nuts, which are manufactured to tighter dimensional tolerances, are given in BS 3692: 2001. Mechanical properties of the bolts are specified in BS EN ISO898: Part 1: 1999.

Strength grade	Yield stress (N/mm^2)	Tensile stress (N/mm^2)
4.6	240	400
8.8	640	800
10.9	900	1000

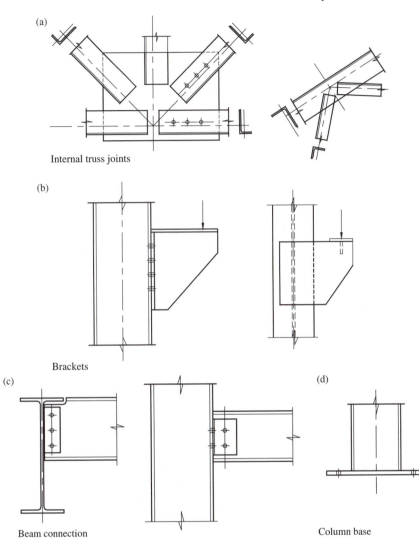

(a)

Internal truss joints

(b)

Brackets

(c)

Beam connection

(d)

Column base

Figure 10.1 Typical connections

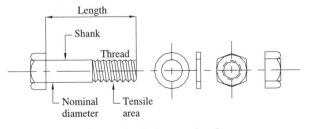

Length

Shank

Thread

Nominal diameter

Tensile area

Hexagon head bolt, nut and washer

Figure 10.2 Hexagon head bolt, nut and washer

The main diameters used are:
    10, 12, 16, 20, (22), 24, (27) and 30 mm
The sizes shown in brackets are not preferred.

10.2.2 Direct shear joints

Bolts may be arranged to act in single or double shear, as shown in Figure 10.3. Provisions governing spacing, edge and end distances are set out in Section 6.2 of BS 5950: Part 1. The principal provisions in normal conditions are:

(1) the minimum spacing is 2.5 times the bolt diameter;
(2) the maximum spacing in unstiffened plates in the direction of stress is $14t$, where t is the thickness of the thinner plate connected;
(3) the minimum edge and end distance as shown in Figure 10.3 from a rolled, machine-flame cut or plane edge is $1.25D$, where D is the hole diameter. For a sheared, hand flame cut edge or any end is $1.40D$.
(4) The maximum edge distance is $11t\varepsilon$, where $\varepsilon = (275/p_y)^{0.5}$.

The standard dimensions of holes for non-preloaded bolts are now specified in Table 33 of BS 5950: Part 1. It depends on the diameter of bolt and the type of bolt hole. In addition to the usual standard clearance, oversize, short and long slotted holes, kidney-shaped slotted hole is now permitted in the revised code. As in usual practice, larger diameter hole is required as the bolt diameter increases. For example, the diameter of a standard clearance hole will be 22 mm for a 20-mm diameter bolt, and 33 mm for a 30-mm diameter bolt.

A shear joint can fail in the following four ways:

(1) by shear on the bolt shank;
(2) by bearing on the member or bolt;
(3) by shear at the end of the member; and
(4) by tension in the member.

The typical failure modes in a shear joint are shown in Figure 10.4(a). These failures modes can be prevented by taking the following measures:

(1) For modes 1 and 2, provide sufficient bolts of suitable diameter.
(2) Provide sufficient end distance for mode 3.
(3) For mode 4, design tension members for effective area (see Chapter 6).

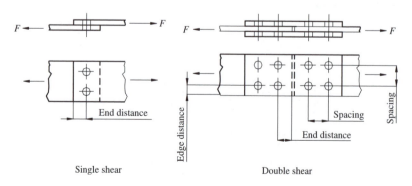

Figure 10.3 Bolts in single and double shear

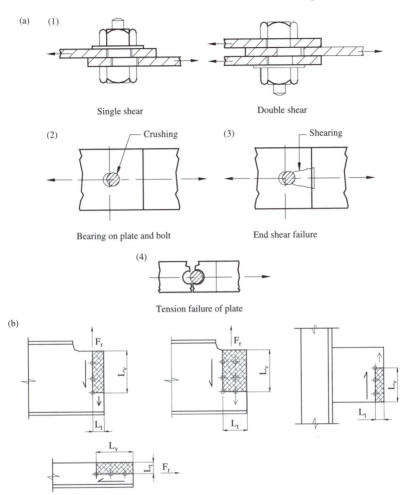

Figure 10.4 (a) Failure modes of a bolted joint (b) Block shear failure

In addition, a new failure mode, block shear, has been observed in a shear joint involving a group of bolts as shown in Figure 10.4(b), and a check of the effective shear area against this failure mode is now required in the revised code.

Also, the revised code now recognises the beneficial effect of strain hardening, and permits the effect of bolt holes on the plate shear capacity to be ignored. See clause 6.2.3 of the revised code for more details.

The design of bolted shear joints is set out in Section 6.3 of BS 5950: Part 1. The basic provisions are:

(1) Effective area resisting shear A_s.
   When the shear plane occurs in the threaded portion of the bolt:

$$A_s = A_t$$

   where A_t is the nominal tensile stress area of the bolt.

When the shear plane occurs in the non-threaded portion:

$$A_s = A$$

where A is the bolt shank area based on the nominal diameter. For a more conservative design, the tensile stress area A_t may be used throughout.

(2) Shear capacity P_s of a bolt:

$$P_s = p_s A_s$$

where p_s is shear strength given in Table 30 of the revised code (see Table 10.1).

It is necessary to reduce the bolt shear capacity in instances where long shear joint and large bolt grip length are used. The revised code now distinguishes between large grip length due to thick component plates and packing plates, and also specifies reduction when kidney-shaped slotted holes are used. See clause 6.3.2 of the revised code for further details.

(3) Block shear:

Block shear failure should also be checked to prevent shear failure through a group of bolts holes at a free edge. The combined block shear capacity for both the shear and tension edges or faces in a shear joint (see Figure 10.4(b)) is given by:

$$P_r = 0.6 p_y t [L_v + k_e (L_t - k D_t)] \geq F_r$$

where D_t is the hole diameter along tension face, k_e the coefficient with values as follows: -0.5 for single lines of bolts; -1.0 for two lines of bolts; L_t the length of tension face, L_v the length of shear face and t the thickness.

(4) Bearing capacity should be taken as lesser of:

Capacity of the bolt, $P_{bb} = d t_p p_{bb}$

where d is the nominal diameter of bolt, t_p the thickness of connected part and p_{bb} the bearing strength of bolt given in Table 31 of the code (see Table 10.1).

Table 10.1 Non-preloaded bolts in standard clearance holes (shear and bearing strengths of bolts and connected parts in N/mm^2)

Strength of bolts	Bolt grade					
	4.6	8.8	10.9	S275[a]	S355[a]	S460[a]
Shear strength p_s	160	375	400	–	–	–
Bearing strength p_{bb}	460	1000	1300	–	–	–
Bearing strength p_{bs}	–	–	–	460[b]	550[b]	670[b]

[a] Steel grade.
[b] Connected parts.

Capacity of the connected part:

$$P_{bs} = k_{bs} d t_p p_{bs} \leq 0.5 k_{bs} e_t p_{bs}$$

where p_{bs} is the bearing strength of the connected parts given in Table 32 of the code (see Table 10.1), e the end distance and k_{bs} the coefficient depending on the type of hole: 1.0 for standard clearance hole, 0.7 for oversize, short or long slotted hole, and 0.5 for kidney-shaped slotted hole.

The second part of the bearing check ensures that the plate does not fail by end shear as shown in Figure 10.4(a) – mode 3.

Load-capacity tables can be made up for ease in design and Table 10.2 is an example. Such tables can be found in the SCI Publication 202: Steelwork Design Guide to BS 5950: Part 1: 2000, Volume 1 Section Properties and Member Capacities, 6th edition with Amendments, The Steel Construction Institute, UK.

With regard to Table 10.2, it should be noted that the minimum end distance to ensure that the bearing capacity of the connected part is controlled by the bearing on the plate, and is given by equating:

$$P_{bs} = k_{bs} d t_p p_{bs} = 0.5 k_{bs} e_t p_{bs}$$

Hence, end distance, $e = 2d$.

10.2.3 Direct tension joints

Two methods are now permitted; either the simple or more exact method can be used. The simple method covers the prying action (see Figure 10.5(b)) by a reduced bolt strength (as before), whereas the more exact method uses the full bolt tension capacity where prying is zero or allowed for in the applied load.

(1) Tension capacity of bolts—simple method

An example of a joint with bolts in direct tension is shown in Figure 10.5(a). The tension capacity using the simple method is:

$$P_t = P_{nom} 0.8 p_t A_t$$

where p_t is the tension strength from Table 34 of the code, $= 240\,\text{N/mm}^2$ for Grade 4.6 bolts, $= 560\,\text{N/mm}^2$ for Grade 8.8 bolts and A_t the nominal tensile stress area.

In this method, the prying force need not be calculated; however, the revised code now places some limitations on the cross-centre spacing of the bolt lines and double curvature bending of the connected end plate.

(2) Tension capacity of bolts—more exact method

Using the more exact method, the tension capacity is:

$$P_t = p_t A_t$$

Prying forces are set up in the T-joint with bolts in tension, as shown in Figure 10.5(b). This situation occurs in many standard joints in structures such as in bracket and moment connections to columns.

Table 10.2 Load capacity table (ordinary non-preloaded bolts Grade 4.6 in S275 steel)

Diameter of Bolt	Tensile Stress Area	Tension Capacity		Shear Capacity		Bearing Capacity in kN (Minimum of P_{bb} and P_{bs}) End distance equal to $2 \times$ bolt diameter										
		Nominal $0.8A_tP_t$ P_{nom}	Exact A_tP_t P_t	Single Shear P_s	Double Shear $2P_s$	Thickness in mm of ply passed through										
	A_t					5	6	7	8	9	10	12	15	20	25	30
mm	mm²	kN	kN	kN	kN											
12	84.3	16.2	20.2	13.5	27.0	27.6	33.1	38.6	44.2	49.7	55.2	66.2	82.8	110	138	166
16	157	30.1	37.7	25.1	50.2	36.8	44.2	51.5	58.9	66.2	73.6	88.3	110	147	184	221
20	245	47.0	58.8	39.2	78.4	46.0	55.2	64.4	73.6	82.8	92.0	110	138	184	230	276
22	303	58.2	72.7	48.5	97.0	50.6	60.7	70.8	81.0	91.1	101	121	152	202	253	304
24	353	67.8	84.7	56.5	113	**55.2**	66.2	77.3	88.3	99.4	110	132	166	221	276	331
27	459	88.1	110	73.4	147	**62.1**	74.5	86.9	99.4	112	124	149	186	248	311	373
30	561	108	135	89.8	180	**69.0**	**82.8**	96.6	110	124	138	166	207	276	345	414

Values in **bold** are less than the single shear capacity of the bolt.

Values in *italic* are greater than the double shear capacity of the bolt.

Bearing values assume standard clearance holes.

If oversize or short slotted holes are used, bearing values should be multiplied by 0.7.

If long slotted or kidney shaped holes are used, bearing values should be multiplied by 0.5.

If appropriate, shear capacity must be reduced for large packings, large grip lengths and long joints.

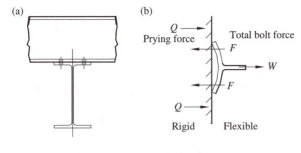

Bolts in tension   Prying force

Figure 10.5 Bolts in tension

The prying force Q adds directly to the tension in the bolt. Referring to the figure,

Total bolt tension, $F = W/2 + Q \leq P_t$

where W is the external tension on the joint.

In the more exact method, it is necessary to calculate the prying force. The magnitude of the prying force depends on the stiffnesses of the bolt and the flanges. Theoretical analyses based on elastic and plastic theory are available to determine the values of these prying forces. Readers should consult further specialised references for this purpose. If the flanges are relatively thick, the bolt spacing not excessive and the edge distance sufficiently large, the prying forces are small and may be neglected.

Where possible, it is recommended that the simple method which covers the prying action using a reduced bolt strength be used because the computation of the actual prying force may be laborious, and this method is adopted here.

10.2.4 Eccentric connections

There are two principal types of eccentrically loaded connections:

(1) Bolt group in direct shear and torsion; and
(2) Bolt group in direct shear and tension.

These connections are shown in Figure 10.6.

10.2.5 Bolts in direct shear and torsion

In the connection shown in Figure 10.6(a), the moment is applied in the plane of the connection and the bolt group rotates about its centre of gravity. A linear variation of loading due to moment is assumed, with the bolt furthest from the centre of gravity of the group carrying the greatest load. The direct shear is divided equally between the bolts and the side plates are assumed to be rigid.

Consider the group of bolts shown in Figure 10.7(a), where the load P is applied at an eccentricity e. The bolts A, B, etc. are at distances r_1, r_2, etc. from the centroid of the group. The coordinates of each bolt are (x_1, y_1), (x_2, y_2), etc. Let the force due to the moment on bolt A be F_T. This is the force on the bolt farthest from the centre of rotation. Then the force on a bolt r_2 from the

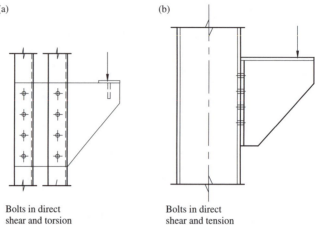

Bolts in direct shear and torsion

Bolts in direct shear and tension

Figure 10.6 Eccentrically loaded connections

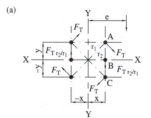

Bolt group in direct shear and torsion

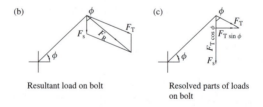

Resultant load on bolt

Resolved parts of loads on bolt

Figure 10.7 Bolt group in direct shear and torsion

centre of rotation is $F_T r_2/r_1$ and so on for all the other bolts in the group. The moment of resistance of the bolt group is given by Figure 10.7:

The load F_T due to moment on the maximum loaded bolt A is given by

$$F_T = \frac{P \cdot e \cdot r_1}{\sum x^2 + \sum y^2}$$

The load F_S due to direct shear is given by

$$F_S = \frac{P}{\text{No. of bolts}}$$

The resultant load F_R on bolt A can be found graphically, as shown in Figure 10.7(b). The algebraic formula can be derived by referring to Figure 10.7(c).

Resolve the load F_T vertically and horizontally to give

Vertical load on bolt $A = F_s + F_T \cos \phi$
Horizontal load on bolt $A = F_T \sin \phi$
Resultant load on bolt A

$$F_R = [(F_T \sin \phi)^2 + (F_s + F_T \cos \phi)^2]^{0.5},$$
$$= [F_s^2 + F_T^2 + 2F_s F_T \cos \phi)]^{0.5}.$$

The size of bolt required can then be determined from the maximum load on the bolt.

10.2.6 Bolts in direct shear and tension

In the bracket-type connection shown in Figure 10.6(b) the bolts are in combined shear and tension. BS 5950: Part I gives the design procedure for these bolts in Clause 6.3.4.4. This is:

The factored applied shear F_S must not exceed the shear capacity P_s, where $P_s = p_s A_s$. The bearing capacity checks must also be satisfactory. The factored applied tension F_T must not exceed the tension capacity P_T, where $P_T = 0.8 p_t A_t$.

In addition to the above the following relationship must be satisfied:

$$\frac{F_S}{P_S} + \frac{F_T}{P_T} \leq 1.4.$$

The interaction diagram for this expression is shown on Figure 10.8. The shear F_s and tension F_T are found from an analysis of the joint.

An approximate method of analysis that gives conservative results is described first. A bracket subjected to a factored load P at an eccentricity

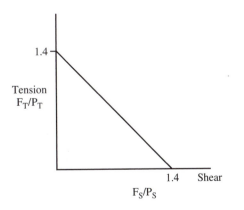

Figure 10.8 Interaction diagram: bolts in shear and tension

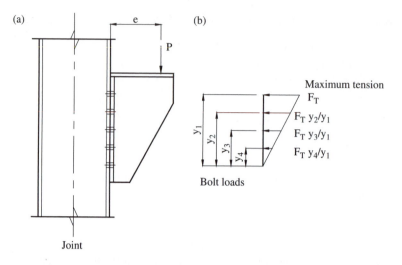

Figure 10.9 Bolts in direct shear and tension: approximate analysis

e is shown in Figure 10.9(a). The centre of rotation is assumed to be at the bottom bolt in the group. The loads vary linearly as shown on the figure, with the maximum load F_T in the top bolt.

The moment of resistance of the bolt group is:

$$
\begin{aligned}
M_R &= 2[F_T \cdot y_1 + F_T \cdot y_2^2/y_1 + \cdots] \\
&= 2F_T/y_1 \cdot [y_1^2 + y_2^2 + \cdots] \\
&= \frac{2\,F_T}{y_1} \cdot \sum y^2 \\
&= P \cdot e
\end{aligned}
$$

The maximum bolt tension is:

$$ F_T = P \cdot e \cdot y_1 / 2 \sum y^2 $$

The vertical shear per bolt:

$$ F_s = P/\text{No. of bolts} $$

A bolt size is assumed and checked for combined shear and tension as described above.

In a more accurate method of analysis, the applied moment is assumed to be resisted by the bolts in tension with uniformly varying loads and an area at the bottom of the bracket in compression, as shown in Figure 10.10(b).

For equilibrium, the total tension T must equal the total compression C. Consider the case where the top bolt is at maximum capacity P_t and the bearing

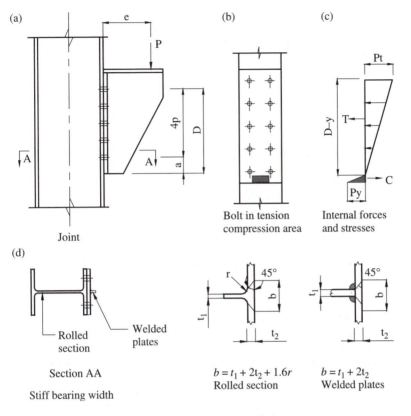

Figure 10.10 Bolts in direct shear and tension: accurate analysis

stress is at its maximum value p_y. Referring to Figure 10.12(c), the total tension is given by:

$$T = P_t + \frac{(D - y - p)}{(D - y)} \cdot P_t + \frac{(D - y - 2p)}{(D - y)} \cdot P_t + \cdots$$

$$P_t = p_t \cdot A_t \text{ (see section 10.2.3(1)).}$$

The total compression is given by:

$$C = 0.5 p_y \cdot b \cdot y$$

where b, the stiff bearing width, is obtained by spreading the load at 45°, as shown in Figure 10.10(d). In the case of the rolled section the 45° line is tangent to the fillet radius. In the case of the welded plates, the contribution of the fillet weld is neglected (for a more accurate stiff bearing length, reader can refer to Clause 4.5.1.3 of the code).

The expressions for T and C can be equated to give a quadratic equation which can be solved to give y, the location of the neutral axis. The moment of resistance is then obtained by taking moments of T and C about the neutral axis. This gives

$$M_R = P_t(D - y) + P_t\frac{(D - y - p)^2}{(D - y)} + \cdots + 2/3 \cdot C \cdot y.$$

The actual maximum bolt tension F_T is then found by proportion, as follows:

Applied moment $M = Pe$,
Actual bolt tension $F_T = MP_t/M_R$.

The direct shear per bolt $F_S = P/$No. of bolts and the bolts are checked for combined tension and shear.

Note that to use the method a bolt size must be selected first and the joint set out and analysed to obtain the forces on the maximum loaded bolt. The bolt can then be checked.

10.2.7 Examples of non-preloaded bolted connections

Example (1)

The joint shown in Figure 10.11 is subjected to a tensile dead load of 85 kN and a tensile imposed load of 95 kN. All data regarding the member and joint are shown in the figure. The steel is Grade S275 and the bolt Grade 4.6. Check that the joint is satisfactory.

Using load factors from Table 2 of BS 5950: Part 1:

Factored load $= (1.4 \times 85) + (1.6 \times 95) = 271$ kN
Strength of bolts from Table 10.2 for 20-mm diameter bolts:
Single shear capacity on threads $= 39.2$ kN
Bearing capacity of bolts on 10-mm ply $= 87.0$ kN
Bearing capacity on 10-mm splice with 30-mm end distance.

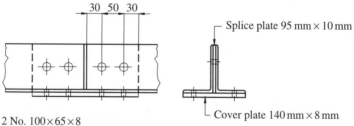

30 50 30

Splice plate 95 mm × 10 mm

Cover plate 140 mm × 8 mm

2 No. 100×65×8
Bolts– 20 mm dia. ordinary non-preloaded bolts
Standard clearance holes– 22 mm dia.

Figure 10.11 Double-angle splice

Bearing strength $p_{bs} = 460\,\text{N/mm}^2$ from Table 10.1:

$$P_{bs} = 1/2 \times 30 \times 10 \times 460/10^3 = 69\,\text{kN}$$

Bolt capacity-two bolts are in double shear and four in single.

$$\text{Shear} = (4 \times 39.2) + (2 \times 39.2) + 69 = 304.2\,\text{kN}$$

Note that the capacity of the end bolt bearing on the 10 mm splice plate is controlled by the end distance (BS 5950: Part 1, Clause 6.3.3.3).

Strength of the angles. Gross area $= 12.7\,\text{cm}^2$ per angle.

The angles are connected through both legs. Clause 4.6.3.3 of BS 5950: Part 1 states that the net area defined in Clause 3.3.2 is to be used in design. The standard clearance holes are 22-mm diameter:

Net area $= 2(1270 - 2 \times 22 \times 8) = 1836\,\text{mm}^2$.
Design strength $p_y = 275\,\text{N/mm}^2$ (Table 9 of code).
Capacity $P_t = 275 \times 1836/10^3 = 504.9\,\text{kN}$.
Splice plate and cover plate (see Clauses 3.3.3 of the code and Section 7.4.1 of this book):
Effective area $= 1.2[(95 - 22)10 + (140 - 44)8] = 1798\,\text{mm}^2$
$\qquad\qquad < \text{gross area} = 2070\,\text{mm}^2$.
Capacity $P_t = 275 \times 1798/10^3 = 494.3\,\text{kN}$.

The splice is adequate to resist the applied load. By inspection, block shear failure is also not critical for this splice joint.

Example (2)

Check that the bracket shown in Figure 10.12 is adequate. All data required are given in the figure.

$$\text{Factored Load} = (1.4 \times 60) + (1.6 \times 80) = 212\,\text{kN}$$
$$\text{Moment } M = (212 \times 525)/10^3 = 111.3\,\text{kN\,m}$$
$$\text{Bolt group } \sum x^2 = 12 \times 250^2 = 750 \times 10^3$$
$$\sum y^2 = 4(35^2 + 105^2 + 175^2) = 171.5 \times 10^3$$
$$\sum x^2 + \sum y^2 = 921.5 \times 10^3$$
$$\cos\phi = 250/305.16 = 0.819$$

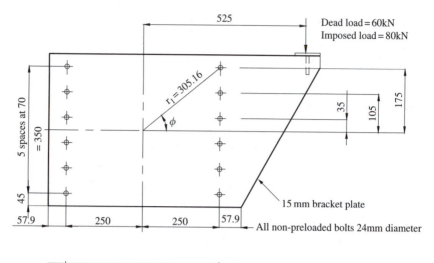

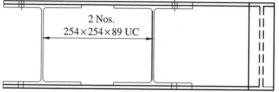

Figure 10.12 Bracket: bolt group in direct shear and torsion

Bolt A is the bolt with the maximum load:

$$\text{Load due to moment} = \frac{111.3 \times 10^3 \times 305.16}{921.5 \times 10^3} = 36.85 \,\text{kN},$$

Load due to direct shear $= 212/12 = 17.67 \,\text{kN}.$

Resultant load on bolt

$$= [17.67^2 + 36.85^2 + (2 \times 17.67 \times 36.85 \times 0.819)]^{0.5}$$
$$= 52.31 \,\text{kN}.$$

Single-shear value of 24 mm diameter non-preloaded bolt on the thread, standard tolerance hole:

From Table 10.2 $= 56.5 \,\text{kN}$,
Universal column flange thickness $= 17.3 \,\text{mm}$,
Side-plate thickness $= 15 \,\text{mm}$,
Minimum end distance $= 45 \,\text{mm}$,
Bearing capacity of the bolt
$P_{bb} = 24 \times 15 \times 460/10^3 = 165.6 \,\text{kN},$

Bearing capacity of the plate
$$P_{bs} = 24 \times 15 \times 460/10^3 = 166 \, \text{kN},$$
$$\le 1/2 \times 45 \times 15 \times 460/10^3 = 155 \, \text{kN}.$$

The strength of the joint is controlled by the single shear value of the bolt. The joint is satisfactory.

Example (3)

Determine the diameter of ordinary non-preloaded bolts required for the bracket shown in Figure 10.13. The joint dimensions and loads are shown in the figure. Use Grade 4.6 bolts, standard clearance holes.

Try 20-mm diameter non-preloaded bolts. From Table 10.2,
Tension capacity $P_T = 47.0 \, \text{kN}$
The design strength from Table 6 of the code for plates:
20 mm thick $p_y = 265 \, \text{N/mm}^2$

Referring to Figure 10.13(c), where the depth to the neutral axis is the total tension is:

$$T = 94\left[1 + \frac{(280 - y)}{(350 - y)} + \frac{(210 - y)}{(350 - y)} + \frac{(140 - y)}{(350 - y)} + \frac{(70 - y)}{(350 - y)}\right]$$
$$= 94[(1050 - 5y)/(350 - y)] \, \text{kN}$$

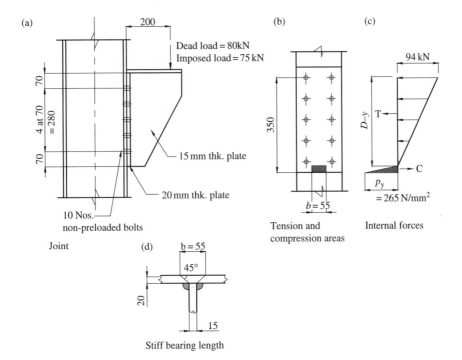

Figure 10.13 Bracket: bolts in direct shear and tension

The width of stiff bearing is shown in Figure 10.13(d):

$$b = 15 + 2 \times 20 = 55\,\text{mm}.$$

The total compression in terms of y is given by:

$$C = \tfrac{1}{2} \times 265 \times 55y/10^3 = 7.29\,\text{kN}.$$

Equate T and C and rearrange terms to give the quadratic equation:

$$7.29y^2 - 3028y + 100\,170 = 0.$$

Solve to give

$$y = 36.25\,\text{mm}.$$

The moment of resistance is:

$$
\begin{aligned}
M_R &= \frac{94}{10^3}\left((350 - 36.5) + \frac{(280 - 36.5)^2}{(350 - 36.25)} + \frac{(210 - 36.5)^2}{(350 - 36.25)}\right. \\
&\quad \left. + \frac{(140 - 36.5)^2}{(350 - 36.25)} + \frac{(70 - 36.5)^2}{(350 - 36.25)}\right) + \frac{7.29 \times 36.5^2 \times 2}{3 \times 10^3} \\
&= 67.17\,\text{kN m}.
\end{aligned}
$$

Factored load $= (1.4 \times 80) + (1.6 \times 75) = 232\,\text{kN}$,
Factored moment $= 232 \times 200/10^3 = 46.4\,\text{kN m}$,
Actual tension in top bolts
$F_T = 46.4 \times 47.0/67.17 = 32.5\,\text{kN}$,
Direct shear per bolt $= 232/10 = 23.2\,\text{kN}$,
Shear capacity on threads $P_S = 39.2\,\text{kN}$,
Tension capacity $P_T = 47.0\,\text{kN}$.

Combined shear and tension:

$$\frac{F_S}{P_S} + \frac{F_T}{P_T} = \frac{23.2}{39.2} + \frac{32.5}{47.0} = 1.28 < 1.4.$$

Therefore 20-mm diameter bolts are satisfactory.

The reader can redesign the bolts using the approximate method. Note that only the bolts have been designed. The welds and bracket plates must be designed and the column checked for the bracket forces. These considerations are dealt with in Section 10.4.5.

10.3 Preloaded bolts

10.3.1 General considerations

Preloaded friction-grip bolts are made from high-strength steel so they can be tightened to give a high shank tension. The shear in the connected plates is transmitted by friction as shown in Figure 10.14 and not by bolt shear, as in ordinary non-preloaded bolts. These bolts are used where strong joints are required, and a major use is in the joints of rigid continuous frames.

The bolts are manufactured in three types conforming to BS4395:

General grade: The strength is similar to Grade 8.8 ordinary non-preloaded bolts. This type is generally used.

Higher-grade: Parallel shank and waisted shank bolts are manufactured in this grade.

The use of general grade preloaded bolts in structural steelwork is specified in BS 4604, Part 1. Two types of preloaded bolts, i.e. parallel and waisted shank bolts are shown in Figure 10.15. The parallel shank bolts are more commonly used these days. Only general grade preloaded bolts will be discussed here.

The bolts must be used with hardened steel washers to prevent damage to the connected parts. The surfaces in contact must be free of mill scale, rust, paint, grease, etc. which would prevent solid contact between the surfaces and lower the slip factor (see below). Care must be taken to ensure that bolts are tightened up to the required tension, otherwise slip will occur and the joint will then act as an ordinary non-preloaded bolted joint. Methods used to achieve the correct shank tension are:

(1) Part-turning. The nut is tightened up and then forced a further half to three quarters of a turn, depending on the bolt length and diameter.

Shank tension

Friction at interface

Action in preloaded bolt

Figure 10.14 Friction in Preloaded bolts

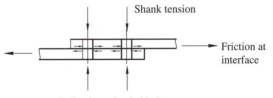

Washer

Waist   Washer

Parallel shank   Waisted shank

Figure 10.15 Types of preloaded bolts

(2) Torque control. A power operated or hand-torque wrench is used to deliver a specified torque to the nut. Power wrenches must be calibrated at regular intervals.
(3) Load-indicating washers and bolts. These have projections which squash down as the bolt is tightened. A feeler gauge is used to measure when the gap has reached the required size.

Friction-grip bolts are generally used in standard clearance holes. The clearances are the same as for ordinary non-preloaded bolts given in Section 10.2.2.

10.3.2 Design procedure

Preloaded bolts can be used in shear, tension and combined shear and tension. The design procedure, given in Section 6.4 of BS 5950: Part 1 for general-grade parallel shank bolts is set out below.

(1) Bolts in shear

The shear capacity is usually the lesser of the slip resistance and the bearing capacity. However, the revised code now provides slip resistance rules based on either serviceability or ultimate criterion. Slip resistances can be calculated for 2 limiting cases, i.e. non-slip in service and non-slip under factored loads. For preloaded bolt based on the serviceability criterion of non-slip in service, a higher slip resistance value can be obtained. However, the bolt can slip and go into bearing at loads greater than the working load; hence its bearing capacity must be checked. On the other hand, if the slip resistance is calculated based on the lower value of non-slip under factored loads, it is not necessary to check for bearing capacity because bearing will not govern the design. This approach is simpler, and is adopted here.

The slip resistance, based on non-slip under factored loads, is given by:

$$P_{SL} = 0.9 K_S \mu P_o$$

where P_o = minimum shank tension (i.e. the proof load given in Table 10.3); $K_S = 1.0$ for bolts in standard clearance holes, and lower values for bolts in oversized and slotted holes (see Clause 6.4.2 of the code); and μ = slip factor. For general-grade preloaded bolts with Class A surface treatment $\mu = 0.5$, other values are given in Table 35 of the code.

The slip resistances for general grade preloaded bolts for various slip factors are given in Table 10.3. The factor K_s is taken as 1.0 for bolts in standard clearance holes.

If bearing resistance is required, it is given by:

$$P_{bg} = 1.5 d t_p p_{bs}$$
$$\leq 0.5 e t_p p_{bs}$$

where d = nominal diameter of the bolt, t_p = thickness of the connected plate, e = end distance, and p_{bs} = bearing strength of the parts connected.

Table 10.3 Load capacities for preloaded bolts: Non-slip under factored loads

Diameter of Bolt	Min. Shank Tension	Bolt Tension Capacity	Slip resistance P_{SL}							
			$\mu = 0.2$		$\mu = 0.3$		$\mu = 0.4$		$\mu = 0.5$	
	P_o	$0.9P_o$	Single Shear	Double Shear	Single Shear	Double Shear	Single Shear	Double Shear	Single Shear	Double Shear
mm	kN	kN	kN	kN	kN	kN	kN	kN	kN	kN
12	49.4	44.5	8.89	17.8	13.3	26.7	17.8	35.6	22.2	44.5
16	92.1	82.9	16.6	33.2	24.9	49.7	33.2	66.3	41.4	82.9
20	144	130	25.9	51.8	38.9	77.8	51.8	104	64.8	130
22	177	159	31.9	63.7	47.8	95.6	63.7	127	79.7	159
24	207	186	37.3	74.5	55.9	112	74.5	149	93.2	186
27	234	211	42.1	84.2	63.2	126	84.2	168	105	211
30	286	257	51.5	103	77.2	154	103	206	129	257

(2) Bolts in tension (non-slip under factored loads)

The tension capacity is given by $P_t = 0.9P_o$
The tension capacities are given in Table 10.3.

(3) Bolts in combined shear and tension (non-slip under factored loads)

The external capacity reduces the clamping action and slip resistance of the joint. Bolts in combined shear and tension must satisfy the following conditions:

(1) The slip resistance must be greater than the applied factored shear F_S
(2) The tension capacity must be greater than the applied factored tension F_{tot}
(3) In addition, the interaction relationship must be satisfied:

$$\frac{F_S}{P_{SL}} + \frac{F_{tot}}{0.9P_o} \le 1$$

10.3.3 Examples of preloaded bolt connections

Example (1)

Design the bolts for the moment and shear connection between the floor beam and column in a steel frame building as shown in Figure 10.16. The following data are given:

Floor beam	610×229 UB 140
Column	254×254 UC 132
Moment Dead load	$= 180 \, kN \, m$
Imposed load	$= 100 \, kN \, m$
Shear Dead load	$= 300 \, kN$
Imposed load	$= 160 \, kN$

(1) Moment connection

The moment is taken by the flange bolts in tension:

Factored moment $= (1.4 \times 180) + (1.6 \times 100) = 412\,\mathrm{kNm}$
Flange force $= 412/0.594 = 693.6\,\mathrm{kN}$

Provide 24-mm diameter preloaded bolts:

Minimum shank tension $= 207\,\mathrm{kN}$
Tension capacity of four bolts $= 4 \times 0.9 \times 207 = 745.2\,\mathrm{kN}$

The joint is satisfactory for moment. Four bolts are also provided at the bottom of the joint but these are not loaded by the moment in the direction shown.

(2) Shear connection

The shear is resisted by the web bolts:

Factored shear $= (1.4 \times 300) + (1.6 \times 160) = 676\,\mathrm{kN}$

Slip resistance of 8 numbers of 24 mm diameter bolts in standard clearance holes $= 8 \times 0.9 \times 1.0 \times 0.5 \times 207 = 745.2\,\mathrm{kN}$
No need for check for bearing because slip resistance under factored loads is used. Therefore, 24-mm diameter bolts are satisfactory.

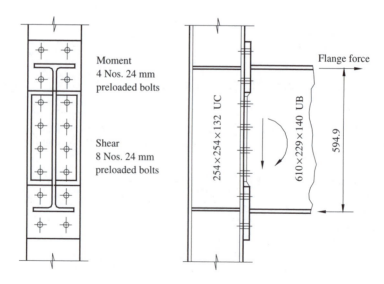

Figure 10.16 Beam-to-column connection

Note that only the bolts have been designed. The welds, end plates and stiffeners must be designed and the column flange and web checked. These considerations are dealt separately in Section 4.5.

Example (2)

Determine the bolt size required for the bracket loaded as shown in Figure 10.17(a).

Factored load $= (1.4 \times 120) + (1.6 \times 110) = 344\,\text{kN}$

Factored moment $= 344 \times 0.28 = 96.3\,\text{kN m}$

Try 20-mm diameter preloaded bolts:

Tension capacity from Table 10.3 $= 130\,\text{kN}$

The joint forces are shown in Figure 10.17(c). The stiff bearing width can be calculated from the bracket end plate (Figure 10.17(d)):

$b = 15 + 2 \times 20 = 55\,\text{mm}$

The total tension in terms of the maximum tension in the top bolts is

$$T = 260\left[1 + \frac{370 - y}{470 - y} + \frac{270 - y}{470 - y} + \frac{170 - y}{470 - y}\right]$$
$$= 260(1280 - 4y)/(470 - y)$$

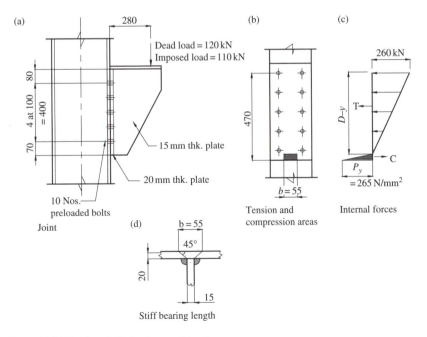

Figure 10.17 Bracket: bolts in shear and tension

The design strength from Table 9of the code for plates 20 mm thick, $p_y = 265$ N/mm^2

The total compression is given by:

$$C = 1/2 \times 265 \times 55y/10^3$$

Equate T and C and rearrange to give:

$$y^2 - 631.9y + 51832 = 0$$

Solving gives $y = 96.87$ mm
The moment of resistance is

$$MR = \frac{260}{10^2} \left((470 - 96.87) + \frac{(370 - 96.87)^2}{(470 - 96.87)} + \frac{(270 - 96.87)^2}{(470 - 96.87)} \right.$$
$$\left. + \frac{(170 - 96.87)^2}{(470 - 96.87)^2} \right) + \frac{7.29 \times 96.87^2 \times 2}{3 \times 10^3}$$
$$= 173.07 + 45.61$$
$$= 218.68 \text{ kNm}$$

The actual maximum tension in the top bolts (assuming no prying force) is:

$$F_{\text{tot}} = 96.3 \times 130/218.68 = 57.2 \text{ kN}$$

The slip resistance from Table 10.3 for slip factor of 0.5, $P_{SL} = 68.4$ kN

Applied shear per bolt $F_S = 344/10 = 34$ kN.

$$\frac{F_S}{P_{SL}} + \frac{F_{\text{tot}}}{0.9P_o} = \frac{34}{68.4} + \frac{57.2}{0.9 \times 144} = 0.94 < 1.0$$

Interaction criteria:
Therefore 20-mm diameter bolts are satisfactory.

10.4 Welded connections

10.4.1 Welding

Welding is the process of joining metal parts by fusing them and filling in with molten metal from the electrode. The method is used extensively to join parts and members, attach cleats, stiffeners, end plates, etc. and to fabricate complete elements such as plate girders. Welding produces neat, strong and more efficient joints than are possible with bolting. However, it should be carried out under close supervision, and this is possible in the fabrication shop. Site joints are usually bolted. Though site welding can be done it is costly, and defects are more likely to occur.

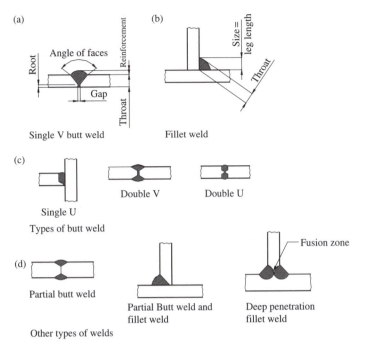

Figure 10.18 Weld types: fillet and butt welds

Electric arc welding is the main system used, and the two main processes in structural steel welding are:

(1) Manual arc welding, using a hand-held electrode coated with a flux which melts and protects the molten metal. The weld quality depends very much on the skill of the welder.
(2) Automatic arc welding. A continuous wire electrode is fed to the weld pool. The wire may be coated with flux or the flux can be supplied from a hopper. In another process an inert gas is blown over the weld to give protection.

10.4.2 Types of welds, defects and testing

The two main types of welds; butt and fillet, are shown in Figures 10.18(a) and (b). Butt welds are named after the edge preparation used. Single and double U and V welds are shown in Figure 10.18(c). The double U welds require less weld metal than the V types. A 90-degree fillet weld is shown but other angles are used. The weld size is specified by the leg length. Some other types of welds – the partial butt, partial butt and fillet weld and deep penetration fillet weld are shown in Figure 10.18(d). In the deep penetration fillet weld a higher current is necessary using submerged arc welding or similar processes to fuse the plates beyond the limit of the weld metal.

Cracks can occur in welds and adjacent parts of the members being joined. The main types are shown in Figure 10.19(a). Contraction on cooling causes cracking in the weld. Hydrogen absorption is the main cause of

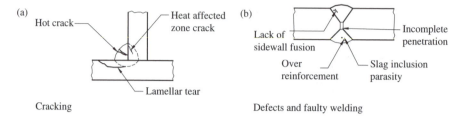

Figure 10.19 Cracks and defects in welds

hydrogen-induced cracking in the heat-affected zone while lamellar tearing along a slag inclusion is the main problem in plates.

Faulty welding procedure can lead to the following defects in the welds, all of which reduce the strength (see Figure 10.19(b)):

(1) Over-reinforcement and undercutting;
(2) Incomplete penetration and lack of side-wall fusion;
(3) Slag inclusions and porosity.

When the weld metal cools and solidifies it contracts and sets up residual stresses in members. It is not economic to relieve these stresses by heat treatment after fabrication, so allowance is made in design for residual stresses.

Welding also causes distortion, and special precautions have to be taken to ensure that fabricated members are square and free from twisting. Distortion effects can be minimized by good detailing and using correct welding procedure. Presetting, pretending and preheating are used to offset distortion. All welded fabrication must be checked, tested and approved before being accepted. Tests applied to welding are given in reference (15):

(1) Visual inspection for uniformity of weld;
(2) Surface tests for cracks using dyes or magnetic particles;
(3) X-ray and ultrasonic tests to check for defects inside the weld.

Only visual and surface tests can be used on fillet welds. Butt welds can be checked internally, and such tests should be applied to important butt welds in tension.

When different thicknesses of plate are to be joined the thicker plate should be given a taper of 1 in 5 to meet the thinner one. Small fillet welds should not be made across members such as girder flanges in tension, particularly if the member is subjected to fluctuating loads, because this can lead to failure by fatigue or brittle fracture. With correct edge preparation if required, fit-up, electrode selection and a properly controlled welding process, welds are perfectly reliable.

10.4.3 Design of fillet welds

Important provisions regarding fillet welds are set out in Clause 6.7.2 of BS5950: Part 1. Some of these are listed below:

(1) End returns for fillet welds around corners should be at least twice the leg length.
(2) In lap joints the lap length should not be less than four times the thickness of the thinner plate.

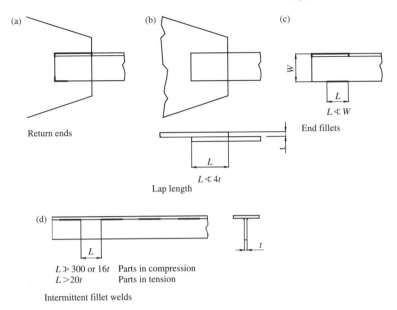

Figure 10.20 Design details for fillet welds

(3) In end connections the length of weld should not be less than the transverse spacing between the welds.

(4) Intermittent welds should not be used under fatigue conditions. The spacing between intermittent welds should not exceed 300 mm or 16t for parts in compression or 24t for parts in tension, where t is the thickness of the thinner plate. These provisions are shown on Figure 10.20.

A key change is the recognition that the fillet weld is stronger in the transverse direction compared to its longitudinal direction, and this has lead to the so-called 'directional method' shown in Figure 10.21. Welds subject to longitudinal shear force is shown in Figure 10.21(a), and those subject to transverse forces are shown in Figure 10.21(b). In a general case, the weld forces have to be analyzed to determine the resultant transverse force on the weld as shown in Figure 10.21(c) to take advantage of this method. It effectively replaces the old symmetrical fillet weld rule which state that the strength of symmetrically loaded fillet welds can be taken as equal to the strength of the plate provided:

(1) The weld strength is not less than that of the plate;
(2) The sum of the throat thicknesses of the weld is greater than the plate thickness; and
(3) The weld is principally in direct compression or tension.

However, for simplicity and ease of use, the code still allows the use of the traditional 'simple method' which does not take into account of the direction of the force acting on the weld. In this method, the vector sum of the design stresses due to all the forces acting on the weld should not exceed its design strength, p_w. Some values of the design strength for fillet welds specified in Table 37, Clause 6.8.5, BS5950: Part I are given here in Table 10.4. This simple method is adopted here; reader to consult the code for the more exact method.

Table 10.4 Design strength of fillet welds p_w (kN/mm^2)

Steel grade BS EN10025	Electrode classification (BS EN499, BS EN440)		
	35	42	50
S275	220	(220)	(220)
S355	(220)	250	(250)
S460	(220)	(250)	280

Note: bracket values are under or over matching electrodes

Table 10.5 Strength of fillet weld (kN/mm run)

Weld size or leg length			Steel grade S355		S460
		S275			
			Electrode classification		
	35	42	42	50	50
5		0.77	0.88	0.88	0.98
6		0.92	1.05	1.05	1.18
8		1.23	1.40	1.40	1.57
10		1.54	1.75	1.75	1.96
12		1.85	2.10	2.10	2.35

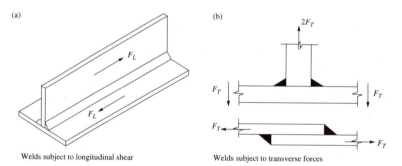

(a) Welds subject to longitudinal shear

(b) Welds subject to transverse forces

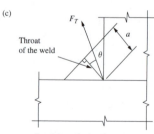

(c) Resultant transverse force on weld

Figure 10.21 Fillet welds—Directional method

In the simple method, the strength of a fillet weld is calculated using the throat thickness. For the 90° fillet weld shown in Figure 10.18(b) the throat thickness is taken as 0.7 times the size or leg length:

Strength of weld = 0.7 leg length $\times p_w/10^3$ kN/mm,
Strengths of fillet weld are given in Table 10.5.

(a)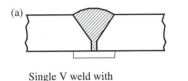

Single V weld with
backing plate

(b)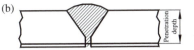

Single V weld made
from one side

Figure 10.22 Design details for butt welds

10.4.4 Design of butt welds

The design of butt welds is covered in Clause 6.9 of BS 5950: Part 1. This clause states that the design strength should be taken as equal to that of the parent metal provided matching electrodes are used. A matching electrode should have specified tensile strength, yield strength, elongation at failure and Charpy impact value each equivalent to, or better, than those specified for the parent metal.

Full penetration depth is ensured if the weld is made from both sides or if backings run is made on a butt weld made from one side (see Figures 10.18(a) and (c)). Full penetration is also achieved by using a backing plate as shown in Figure 10.22(a). If the weld is made from one side the throat thickness may be reduced. The throat size of a partial penetration butt weld should be taken as the minimum depth of penetration from that side of the weld as shown in Figure 10.22(b). The capacity of the partial penetration butt weld should be taken as sufficient if throughout the weld the stress does not exceed the relevant strength of the parent material.

The code also states in Clause 6.9.3 that if the weld is unsymmetrical relative to the parts joined the resulting eccentricity should be taken into account in calculating the stress in the weld.

10.4.5 Eccentric connections

The two types of eccentrically loaded connections are shown in Figure 10.23. These are:

(1) The torsion joint with the load in the plane of the weld; and
(2) The bracket connection.

In both cases, the fillet welds are in shear due to direct load and moment.

10.4.6 Torsion joint with load in plane of weld

The weld is in direct shear and torsion. The eccentric load causes rotation about the centre of gravity of the weld group. The force in the weld due to torsion is taken to be directly proportional to the distance from the centre of gravity and is found by a torsion formula. The direct shear is assumed to be uniform throughout the weld. The resultant shear is found by combining the shear due to moment and the direct shear, and the procedure is set out below. The side plate is assumed to be rigid.

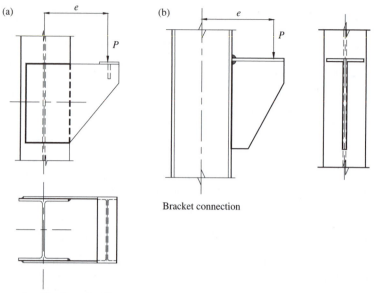

Bracket connection

Load in plane of weld

Figure 10.23 Eccentrically loaded connections

A rectangular weld group is shown in Figure 10.24(a), where the eccentric load P is taken on one plate. The weld is of unit leg length throughout:

Direct shear $F_s = P/\text{length of weld}$
$$= P/[2(x + y)].$$

Shear due to torsion

$$F_T = Per/I_p$$

where,

I_p is the polar moment of inertia of the weld group $= I_x + I_y$,
$I_x = (y^3/6) + (xy^2/2)$,
$I_y = (x^3/6) + (x^2y/2)$,
$r = 0.5(x^2 + y^2)^{0.5}$.

The heaviest loaded length of weld is that at A, furthest from the centre of rotation O. The resultant shear on a unit length of weld at A is given by:

$$F_R = [F_S^2 + F_T^2 + 2F_S F_T \cos\phi]^{0.5}$$

The resultant shear is shown on Figure 10.24(b). The weld size can be selected from Table 10.5.

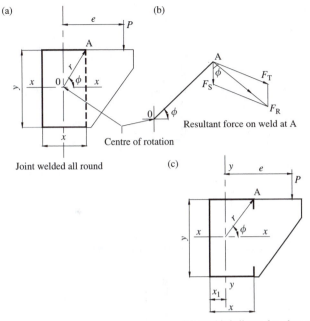

(a) Joint welded all round

(b) Resultant force on weld at A

(c) Joint welded all round on three side

Figure 10.24 Torsion joints load in plane of weld

If the weld is made on three sides only, as shown on Figure 10.24(c), the centre of gravity of the group is found first by taking moments about side BC:

$$x_1 = x^2/(2x + y)$$
$$I_x = y^3/12 + xy^2/2$$
$$I_y = x^3/6 + 2x(x/2 - x_1)^2 + yx_1^2$$

The above procedure can then be applied.

10.4.7 Bracket connection

Various assumptions are made for the analysis of forces in bracket connections. Consider the bracket shown in Figure 10.25(a), which is cut from a universal beam with a flange added to the web. The bracket is connected by fillet welds to the column flange. The flange welds have a throat thickness of unity and the web welds a throat thickness q, a fraction of unity. Assume rotation about the centroidal axis XX. Then:

Weld length	$L = 2b + 2aq$,
Moment of inertia	$I_x = bd^2/2 + qa^3/6$,
Direct shear	$F_s = P/L$,
Load due to moment	$F_T = Ped/2I_x$,
Resultant load	$F_R = (F_T^2 + F_S^2)^{0.5}$.

Select the weld size from Table 10.5.

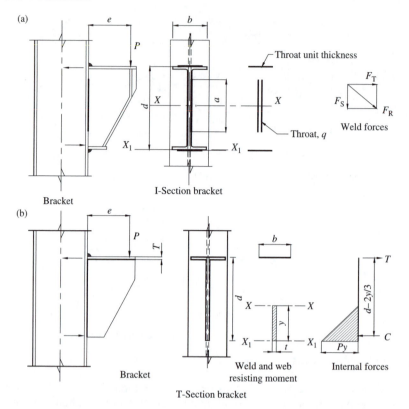

Figure 10.25 Bracket connections

In a second assumption rotation takes place about the bottom flange X_1X_1. The flange welds resist moment and web welds shear. In this case:

$$F_T = Pe/db$$
$$F_s = P/2a.$$

The weld sizes can be selected from Table 10.5.

With heavily loaded brackets full-strength welds are required between the bracket and column flange.

The fabricated T-section bracket is shown in Figure 10.25(b). The moment is assumed to be resisted by the flange weld and a section of the web in compression of depth y, as shown in the figure. Shear is resisted by the web welds.

The bracket and weld dimensions and internal forces resisting moment are shown in the figure. The web area in compression is ty. Equating moments gives:

$$P \cdot e = T(d - 2y/3) = C(d - 2y/3)$$
$$C = \tfrac{1}{2}p_y ty = T$$
$$P \cdot e = \tfrac{1}{2}p_y ty(d - 2y/3).$$

Solve the quadratic equation for y and calculate C:

The flange weld force $F_T = T/b$,
The web weld force $F_s = P/2(d - y)$.

Select welds from Table 10.5.

The calculations are simplified if the bracket is assumed to rotate about the $X_1 X_1$ axis, when

$$F_T = Pe/db.$$

10.4.8 Examples on welded connections

(1) Direct shear connection

Design the fillet weld for the direct shear connection for the angle loaded as shown in Figure 10.26(a), where the load acts through the centroidal axis of the angle. The steel is Grade S275:

Factored load $= (1.4 \times 50) + (1.6 \times 60) = 166$ kN,

Use 6-mm fillet weld, strength from Table 10.5 $= 0.92$ kN/mm,

Length required $= 166/0.92 = 184.4$ mm,

Balance the weld on each side as shown in Figure 10.26(b):

Side X, length $= 184.4 \times 43.9/65 = 124.5$ mm,

   Add 12 mm, final length $= 136.5$ mm, say 140 mm,

Side Y, length $= 184.4 - 124.5 = 59.9$ mm,

   Add 12 mm, final length $= 71.9$, say 75 mm.

Note that the length on side Y exceeds the distance between the welds, as required in Clause 6.7.2.6 of BS 5950: Part 1. A weld may also be placed across the end of the angle, as shown in Figure 10.26(c). The length of weld on side Y,

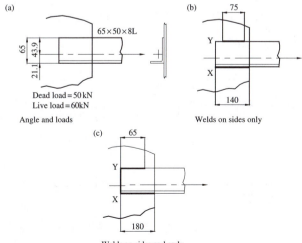

Figure 10.26 Direct shear connection

L_y may be found by taking moments about side X. In terms of weld lengths this gives:

$(L_y \times 65) + (65 \times 32.5) = (184.4 \times 21.1)$
$L_y = 27.4 + 6 = 33.4$, say 40 mm.

Length on side X:
$L_x = 184.4 - 65 - 27.4 + 6 = 98$ mm, say 100 mm.

Note that the leg length has been added at ends of all weld lengths calculated above to allow for craters at the ends. To comply with Clause 6.7.2.6 quoted above, weld L_y is increased to 65 mm. Weld L_x is also increased in proportion to 180 mm. In the above example, the load is also eccentric to the plane of the gusset plate, as shown in Figure 10.26(a). It is customary to neglect this eccentricity.

(2) Torsion connection with load in plane of weld

One side plate of an eccentrically loaded connection is shown in Figure 10.27(a). The plate is welded on three sides only. Find the maximum shear force in the weld and select a suitable fillet weld from Table 10.5.

Find the position of the centre of gravity of the weld group by taking moments about side AB (see Figure 10.27(b)):

Length $L = 700$ mm,
Distance to centroid $x_1 = 2 \times 200 \times 100/700 = 57.14$ mm,
Eccentricity of load $e = 292.86$.

Moments of inertia:

$$I_x = (2 \times 200 \times 150^2) + 300^3/12 = 11.25 \times 10^6 \text{ mm}^3,$$

$$I_y = (300 \times 57.14^2) + (2 \times 200^3/12) + (2 \times 200 \times 42.86^2)$$
$$= 3.047 \times 10^6 \text{ mm}^3,$$

$$I_p = (11.25 + 3.047)10^6 = 14.297 \times 10^6 \text{ mm}^2,$$

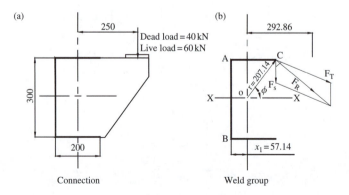

Figure 10.27 Torsion connection loaded in plane of weld

Angle $\cos \phi = 142.86/207.14 = 0.689$
Factored load $= (1.4 \times 40) + (1.6 \times 60) = 152\,\text{kN}$,
Direct shear $F_\text{s} = 152/700 = 0.217\,\text{kN/mm}$.

Shear due to torsion on weld at C:

$$F_\text{T} = \frac{152 \times 292.86 \times 207.14}{14.297 \times 10^6} = 0.645\,\text{kN/mm}.$$

Resultant shear:

$$F_\text{R} = [0.217^2 + 0.645^2 + 2 \times 0.217 \times 0.645 \times 0.689]^{0.5}$$
$$= 0.81\,\text{kN/mm}.$$

A 6-mm fillet weld, strength 0.92 kN/mm is required.

(3) Bracket connection

Determine the size of fillet weld required for the bracket connection shown in Figure 10.28. The web welds are to be taken as one half the leg length of the flange welds. All dimensions and loads are shown in the figure.
Design assuming rotation about XX axis

Factored load $= (1.4 \times 80) + (1.6 \times 110) = 288\,\text{kN}$,

Length $L = (2 \times 173.2) + 280 = 626.4\,\text{mm}$,

Inertia $I_x = (2 \times 173.2 \times 182^2) + 280^3/12 = 13.3 \times 10^6\,\text{mm}^3$,

Direct shear $F_\text{s} = 288/626.4 = 0.46\,\text{kN/mm}$,

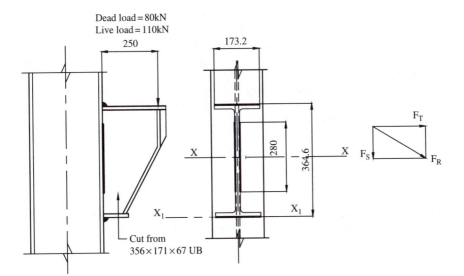

Figure 10.28 Bracket connection

Shear from moment $F_T = \dfrac{288 \times 250 \times 182}{13.3 \times 10^6} = 0.985\,\text{kN/mm}$,

Resultant shear $F_R = [0.462 + 0.985^2]^{0.5} = 1.09\,\text{kN/mm}$.

Provide 8-mm fillet welds for the flanges, strength 1.23 kN/mm. For the web welds provide 6-mm fillets (the minimum size recommended).
Design assuming rotation about X_1X_1 axis
The flange weld resists the moment -288×250

$$F_T = \frac{288 \times 250}{364 \times 173.2} = 1.14\,\text{kN/mm}.$$

Provide 8-mm fillet welds, strength 1.23 kN/mm. The web welds resist the shear:

$$F_s = 288/(2 \times 280) = 0.514\,\text{kN/mm}.$$

Provide 6-mm fillet welds. The methods give the same results.

10.5 Further considerations in design of connections

10.5.1 Load paths and forces

The design of bolts and welds has been considered in the previous sections. Other checks which depend on the way the joint is fabricated are necessary to ensure that it is satisfactory. Consistent load paths through the joint must be adopted.

Consider the brackets shown in Figure 10.29. The design checks required are:

(1) The bolt group (see Section 10.3.3);
(2) The welds between the three plates (see Section 10.4.5);
(3) The bracket plates. These are in tension, bearing, buckling and local bending;
(4) The column in axial load, shear and moment. Local checks on the flange in bending and web in tension at the top and buckling and bearing at the bottom are also required.

10.5.2 Other design checks

Some points regarding the design checks are set out below.

(1) A direct force path is provided by the flange on the bracket in Figure 10.29(a). The flange can be designed to resist force

$$R = (d^2 + e^2)^{0.5} P/d$$

when e is the eccentricity of the load and d the depth of bracket.

(a)

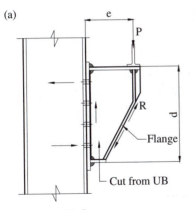

Bracket with flange

(b)

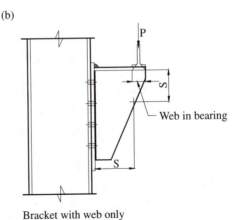

Bracket with web only

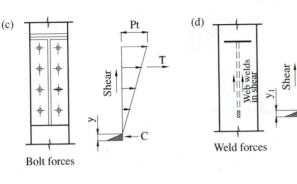

(c) Bolt forces

(d) Weld forces

(e)

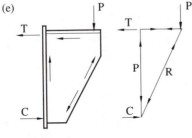

Web plate forces

Figure 10.29 Brackets: load paths and forces

(2) Where the bracket has a web plate only, as shown in Figure 10.29(b), the maximum outstand should not exceed $13te$, where t is the thickness of plate and $\varepsilon = (275/p_y)^{0.5}$.

This ensures that the plate can be stressed to the design strength p_y without buckling. (See Table 9 of BS 5950: Part I.) Local buckling is dealt with in Section 4.3. If the web is satisfactory for bearing under the load P, the load R can be assumed to be carried on a strip of web of width equal to the length in bearing (see (3) below). The bolt, weld and web plate forces for this type of bracket are shown in Figures 10.29(c), (d) and (c), respectively. (See references (16) and (17) for a more rigorous treatment of this problem).

(a)

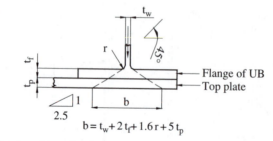

$$b = t_w + 2t_f + 1.6r + 5t_p$$

Bearing at top of web

(b)

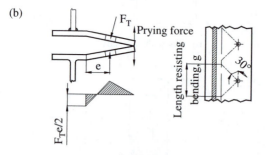

Column flange and end plate in bending

(c)

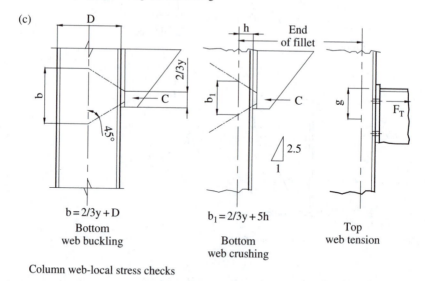

Column web-local stress checks

Figure 10.30 Brackets: local stress check

(3) The bearing length at the top of the web is shown in Figure 10.30(a). For a universal beam, the load is dispersed at 45° from the beam fillets and at 1 in 2.5 through the top plate. (See Clauses 4.5.1.3 and 4.5.2 in BS 5950: Part 1. Bearing is dealt with in Section 4.8.2.)

(4) The end plate and column flange are checked for bending, as shown in Figure 10.30(b). The plates are in double curvature, produced by prying

forces which are absorbed by the bolts. The length resisting bending is found by dispersing the load at 30°, as shown in the figure. For a more rigorous approach using yield line analysis, see reference (18).

(5) The following checks are made on the column at the bottom of the bracket. The column web is checked for bearing at the end of the fillet between flange and web with load dispersed at 1 in 2.5 to give the length in bearing b_1, shown in Figure 10.30(c). Note that the compression force is assumed to spread over a depth $2y/3$, where y is the depth of compression area from the bolt force analysis (see Figure 10.29(c)). The column web is checked for buckling at the centre line of the column. The load is dispersed at 45° to give the length b, shown in Figure 10.30(c), which is considered for buckling. (See Section 4.8.1 for details of this design check.) Stiffeners can be added to carry loads if the column web is overstressed.

(6) The column web is checked in tension at the top of the bracket, as shown in Figure 10.30(c). The length in tension is taken as the length g resisting bending, defined in (4) above.

Problems

10.1 A single-shear bolted lap joint (Figure 10.31) is subjected to an ultimate tensile load of 200 kN. Determine a suitable bolt diameter using Grade 4.6 bolts.

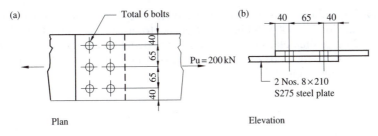

Figure 10.31

10.2 A double-channel member carrying an ultimate tension load of 820 kN is to be spliced, as shown in Figure 10.32.

(1) Determine the number of 20-mm diameter Grade 4.6 bolts required to make the splice.

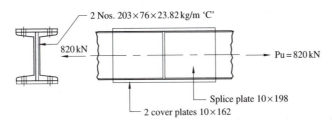

Figure 10.32

(2) Check the double-channel member in tension.
(3) Check the splice plates in tension.

10.3 A bolted eccentric connection (illustrated in Figures 10.33(a) and (b)) is subjected to a vertical ultimate load of 120 kN. Determine the size of Grade 4.6 bolts required if the load is placed at an eccentricity of 300 mm.

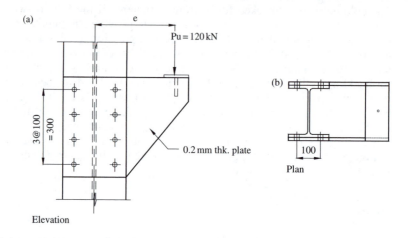

Figure 10.33

10.4 The bolted bracket connection shown in Figure 10.34 carries a vertical ultimate load of 300 KN placed at an eccentricity of 250 mm. Check that 12 No. 24-mm diameter Grade 4.6 bolts are adequate. Use both approximate and accurate methods of analysis discussed in Section 10.2.6. Assume all plates to be 20 mm thick.

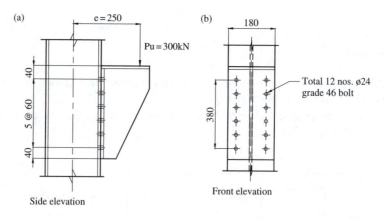

Figure 10.34

10.5 Design a beam–splice connection for a 533 × 210 UB 82. The ultimate moment and shear at the splice are 300 kN m and 175 kN, respectively. A sketch of the suggested arrangement is shown in Figure 10.35. Prepare the final connection detail drawing from your design results.

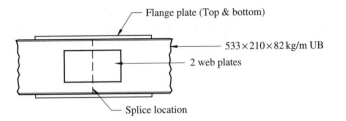

Figure 10.35

10.6 The arrangement for a preloaded bolt grip-connection provided for a tie carrying an ultimate force of 300 kN is shown in Figure 10.36. Check the adequacy if all the bolts provided are 20-mm diameter.

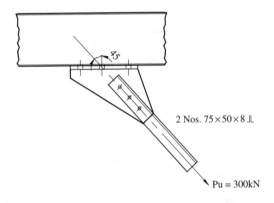

Figure 10.36

10.7 Redesign the bracket connection in Problem 10.4 using high-strength preloaded bolts. What is the minimum bolt diameter required? Discuss the relative merits of using (i) non-preloaded Grade 4.6 bolts, (ii) non-preloaded Grade 8.8 bolts, and (iii) high strength preloaded bolts for connections.

10.8 The welded connection for a tension member in a roof truss is shown in Figure 10.37. Using Class 42 electrode on Grade S275 plate, determine the minimum leg size of the welds if the ultimate tension in the member is 90 kN.

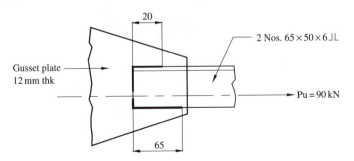

Figure 10.37

10.9 Determine the leg length of fillet weld required for the eccentric joint shown in Figure 10.38. The ultimate vertical load is 500 kN placed at 300 mm from the centre line. Use Class 42 electrode on a Grade S275 plate.

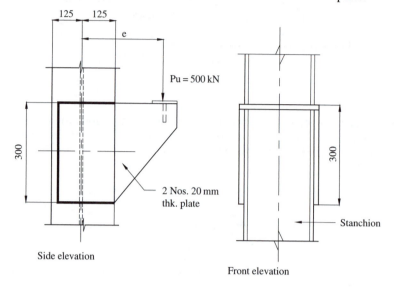

Figure 10.38

10.10 A bracket cut from a 533 × 210 UB 82 of Grade S275 steel is welded to a column, as shown in Figure 10.39. The ultimate vertical load on the bracket is 350 kN applied at an eccentricity of 250 mm. Design the welds between the bracket and column.

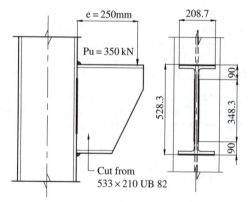

Figure 10.39

11

Workshop steelwork design example

11.1 Introduction

An example giving the design of the steel frame for a workshop is presented here and illustrates the following steps in the design process:

(1) Preliminary considerations and estimation of loads for the various load cases;
(2) Computer analysis for the structural frame;
(3) Design of the truss and crane column;
(4) Sketches of the steelwork details.

The framing plans for the workshop with overhead crane are shown in Figure 11.1. The frames are spaced at 6.0 m centres and the overall length of building is 48.0 m. The crane span is 19.1 m and the capacity is 50 kN.

Design the structure using Grade S275 steel. Structural steel angle sections are used for the roof truss and universal beams for the columns.

Computer analysis is used because plane frame programs are now generally available for use on microcomputers. (The reader can consult references 16 and 22 for particulars of the matrix stiffness method analysis. The plane frame manual for the particular software package used should also be consulted.)

A manual method of analysis could also be used and the procedure is as follows:

(1) The roof truss is taken to be simply supported for analysis.
(2) The columns are analysed for crane and wind loads assuming portal action with no change of slope at the column top. The portal action introduces forces into the chords of the truss which should be added to the forces in (1) for design.

The reader should consult references (9) and (20) for further particulars of the manual method of analysis.

11.2 Basic design loads

Details of sheeting and purlins used are given below:

Sheeting: Cellactite 11/3 corrugated sheeting, type 800 thickness 0.8 mm. Dead load = 0.1 kN/m². The loads and estimated self-weight on

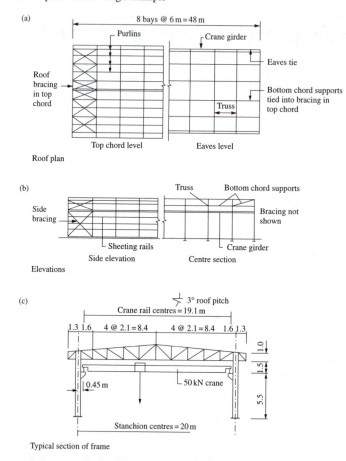

Figure 11.1 Arrangement of workshop steel frame

plan are as follows:

Roof dead load	(kN/m²)
Sheeting	0.11
Insulation and lighting	0.14
Purlin self weight	0.03
Truss and bracing	0.10
Total load on plan	0.38

Imposed load on plan = 0.75 kN/m².

Purlins: Purlins are spaced at 2.1 m centres and span 6.0 m between trusses, Purlin loads = 0.11 + 0.14 + 0.75 = 1.0 kN/m², Provide Ward Building Components Purlin A200/180, Safe load = 1.07 kN/m².

Walls: Cladding, insulation, sheeting rails and bracing = 0.3 kN/m².

Stanchion: Universal beam section, say 457 × 191 UB 67 for self-weight estimation.

Crane data: Hoist capacity $= 50\,\text{kN}$,
             Bridge span $= 19.1\,\text{m}$,
             Weight of bridge $= 35\,\text{kN}$,
             Weight of hoist $= 5\,\text{kN}$,
             End clearance $= 220\,\text{mm}$,
             End-carriage wheel centres $= 2.2\,\text{m}$,
             Minimum hook approach $= 1.0\,\text{m}$.

Wind load: The readers should refer to the latest Code of Practice for wind loads, BS 6399, Part 2.

The structure is located in Northern England, site in country or up to 2 km into town. The basic wind speed V_b is 26 m/s.

11.3 Computer analysis data

11.3.1 Structural geometry and properties

The computer model of the steel frame is shown in Figure 11.2 with numbering for the joints and members. The joint coordinates are shown in Table 11.1. The column bases are taken as fixed at the floor level and the truss joints and connections of the truss to the columns are taken as pinned. The structure is analysed as a plane frame. The steel frame resists horizontal load from wind and crane surge by cantilever action from the fixed based columns and portal action from the truss and columns. The other data and member properties are shown in Table 11.2.

Elastic modulus, $E = 205\,000\,\text{N/mm}^2$

Columns: try 457×191 UB67

$$A = 85.4\,\text{cm}^2, \quad E \times A = 1751\,\text{MN},$$

$$I_x = 29401\,\text{cm}^4 \quad E \times I_x = 60.3\,\text{MN/m}^2.$$

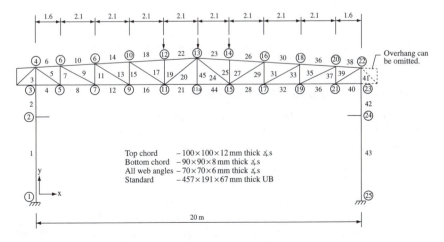

Figure 11.2 Structural model of steel frame

Table 11.1 Joints coordinates of structure

Joint	X-distance	Y-distance	Joint	X-distance	Y-distance
1	0.00	0.000	14	12.1	8.415
2	0.00	5.500	15	12.1	7.000
3	0.00	7.000	16	14.2	8.305
4	0.00	8.000	17	14.2	7.000
5	1.60	7.000	18	16.3	8.195
6	1.60	8.085	19	16.3	7.000
7	3.70	7.000	20	18.4	8.085
8	3.70	8.195	21	18.4	7.000
9	5.80	7.000	22	20.0	8.000
10	5.80	8.305	23	20.0	7.000
11	7.90	7.000	24	20.0	5.500
12	7.90	8.415	25	20.0	0.000
13	10.0	8.525	11a	10.0	7.000

Table 11.2 Structural control data

Number of joints	= 26
Number of members	= 45
Number of joint loaded	= 13
Number of member loaded	= 4

Top chord angles: try $100 \times 100 \times 12\,\text{mm}^3$ *angles*

$$A = 22.7\,\text{cm}^2 \quad E \times A = 465\,\text{MN}.$$

Bottom chord angles: try $90 \times 90 \times 8\,\text{mm}^3$ *angles*

$$A = 13.9\,\text{cm}^2 \quad E \times A = 285\,\text{MN}$$

All web members: try $80 \times 80 \times 6\,\text{mm}$ *angles,*

$$A = 9.35\,\text{cm}^2 \quad E \times A = 192\,\text{MN}.$$

Note: All $E \times I$ *values of the steel angles used in the truss are set to nearly zero.*

11.3.2 Roof truss: dead and imposed loads

For the steel truss the applied loads are considered as concentrated at the purlin node points. With the purlin spacing of 2.1 m, the applied joint loads at the top chord are:

Dead loads per panel point $= 0.38 \times 6 \times 2.1 = 4.8\,\text{kN}$,
Imposed loads per panel point $= 0.75 \times 6 \times 2.1 = 9.45\,\text{kN}$.

Figure 11.3 shows the dead and imposed loads applied on the steel truss at the top chord node points. The dead load of crane girder, side rail, cladding and vertical bracing are assumed to load directly on to the column without affecting the roof truss. The estimated load is 18 kN per leg.

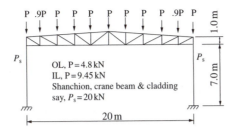

Figure 11.3 Applied dead and imposed loads on truss

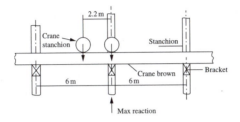

Figure 11.4 Wheels location for maximum reaction

11.3.3 Crane loads

The maximum static wheel load from the manufacturer's table is

Maximum static load per wheel = 35 kN,
Add 25% for impact $35 \times 1.25 = 43.8$ kN.

The location of wheels to obtain the maximum reaction on the column leg is shown in Figure 11.4, with one of the wheels directly over the support point:

Maximum reaction on the column through the crane bracket

$$= 43.8 + 43.8 \times (6 - 2.2)/6$$
$$= 43.8 \times 1.63 = 71.5.$$

Corresponding reaction on the opposite column from the crane

$$= 8.7 \times 1.25 \times 1.63 = 17.7 \text{ kN}.$$

Transverse surge per wheel is 10% of hoist weight plus hook load

$$= 0.1 \times (50 + 5)/4 = 1.4 \text{ kN}.$$

The reaction on the column is

$$= 1.4 \times 71.4/43.8 = 2.28 \text{ kN}.$$

Crane load eccentricity from centre line of column assuming 457×191 UB 67:

$$e = 220 + 457/2 = 448.5 \quad \text{say } 450 \text{ mm}.$$

Figure 11.5 shows the crane loads acting on the frame. The two applied moments are due to the vertical loads multiplied by the eccentricities.

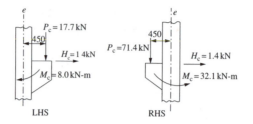

Figure 11.5 Crane loading on steel frame

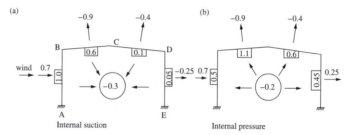

Figure 11.6 Wind-pressure coefficients $\alpha = 0°$

Table 11.3 Wind pressure coefficient on building

External pressure coefficient, C_{pe}		Internal coefficient	
Roof surfaces	Wall surfaces	Suction	Pressure
EF GH	A B		
−0.9 −0.4	+0.7 −0.25	−0.3	+0.2

11.3.4 Wind loads on to the structure

The site wind speed, V_s = basic wind speed $\times S_a \times S_d \times S_s \times S_p$, where S_a is the altitude factor (taken as 1.0), S_d = direction factor (1.0 for all directions) and S_s = seasonal factor and S_p = probability factor, both taken as 1.0.

From Table 4, BS 6399, for effective height H_e of 10 m and closest distance to sea upwind of 10 km, the factor S_b is 1.73.

The effective wind speed $V_e = V_s \times S_b = 26 \times 1.73 = 45$ m/s.

Dynamic wind pressure $q = 0.613 \times 45^2 = 1241$ N/m^2.

The wind pressure coefficient for $\alpha = 0°$ is shown in Figure 11.6 and Table 11.3. The wind forces applied on the frame and used for the computer analysis are given in Figures 11.7(a) and (b) for the internal pressure and suction cases, respectively.

For $\alpha = 90°$, the wind pressure coefficients and calculated wind loads are shown in Figure 11.8.

11.4 Results of computer analysis

A total of seven computer runs were carried out with one run each for the following load cases:

Case 1: Dead loads (DL),

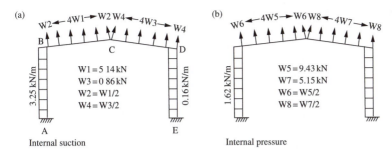

Figure 11.7 Wind loads on structure: α = 0°

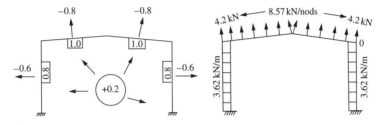

Figure 11.8 Wind loads on structure: α = 90°

Case 2: Imposed loads (IL),
Case 3: Wind loads at 0° angle, internal suction—(WL, IS),
Case 4: Wind loads at 0° angle, internal pressure—(WL, IP),
Case 5: Wind loads at 90° angle, internal pressure—(WL, IM),
Case 6: Crane loads, when maximum wheel loads occur—(CRWL),
Case 7: Crane surge loads—(CRSL).

Table 11.4 gives the summary of the truss member axial forces extracted from the computer output. There are no bending moments in the truss members. It was found that the crane loads do not produce any axial force in truss members listed in Table 11.4, except for those truss members connected directly to the column legs, which have some forces due to crane loads. They are members 3–4, 3–5, 4–5, 4–6, 20–22, 21–22 and 22–23.

Table 11.5 shows the member axial forces and moments for the column legs and the truss members connected directly on to it. There are no bending moments for members 3–4, 3–5, 4–5, 4–6, 20–22, 21–22 and 22–23.

The following five critical loads combinations are computed:

(1) 1.4 DL + 1.6 LL
(2) 1.4 DL + 1.6 LL + 1.6 CRWL
(3) 1.0 DL + 1.4 WL (wind at 0°, I.S.)
(4) 1.0 DL + 1.4 WL (wind at 0°, I.P.)
(5) 1.0 DL + 1.4 WL (wind at 90°, I.P.)

Table 11.4 Summary of member force (kN) for truss

Member	Dead load (DL)	Imposed load (IL)	Wind at T (IP)	Wind at 90* (IP)	1.4 DL + 1.6 IL	DL + 1.4 x wind at 0°,	DL + 1.4 x wind at 90°
4–6	−1.3	−2.56	17.5	3.0	−5.92	23.2	−1.3
6–8	−30.7	−60.45	63	55.6	−139.70	57.5	47.14
8–10	−47.5	−93.53	85.6	85.4	−216.14	72.34	72.06
10–12	−54.5	−107.31	90.7	97.9	−248.00	72.48	82.56
12–13	−54.5	−107.31	90.7	−97.9	−248.00	72.48	82.56
13–14	−54.5	−107.31	90.7	97.9	−248.00	72.48	82.56
14–16	−54.5	−107.31	90.7	97.9	−248.00	72.48	82.56
16–18	−47.5	−93.53	85.6	85.4	−216.14	72.34	72.06
18–20	−30.7	−60.45	63	55.6	−139.70	57.5	47.14
20–22	−1.3	−2.56	17.5	3.0	−5.92	23.2	−1.3
3–5	40.5	79.74	−50.3	−63	184.29	−29.92	−47.7
5–7	6.1	12.01	−6.0	−1.6	27.76	6.1	−38.6
7–9	23.3	45.88	−51.4	−50.9	106.02	−48.66	−47.96
9–11	40	78.76	−74	−80.6	182.02	−63.6	−72.84
11–11a	46.4	91.36	−70.4	−91.9	211.14	−52.16	−82.26
11a–15	46.4	91.36	−70.4	−91.9	211.14	−52.16	−82.26
15–17	40	78.76	−47.3	−80.6	182.02	−26.22	−72.84
17–19	23.3	45.88	−18.9	−50.9	106.02	−3.16	−47.96
19–21	6.1	12.01	−25.1	−1.6	27.76	−29.04	3.86
21–23	40.5	79.74	−73.4	−63	184.29	−62.26	−47.7
4–5	40.6	79.94	−66.4	−72.3	184.75	−52.36	−60.62
5–6	−21.5	−42.33	35.2	38.4	−97.83	27.78	32.26
4–5	40.6	79.94	−66.4	−72.3	184.75	−52.36	−60.62
5–6	−21.5	−42.33	35.2	38.4	−97.83	27.78	32.26
6–7	33.1	65.17	−51	−59	150.62	−38.3	−49.5
7–8	−15.2	−29.93	23.5	27.1	−69.17	17.7	22.74
8–9	19.3	38.00	−25.9	−34.3	87.82	−16.96	−28.72
9–10	−9.6	−18.90	12.8	16.9	−43.68	8.32	14.06
10–11	8.3	16.34	6.0	−14.7	−37.77	8.3	−12.28
11–12	−4.8	−9.45	9.4	8.6	−21.84	8.36	7.24
11–13	0.8	9.45	−10.7	−1.4	3.64	−14.18	−1.16
13–11a	0	0	0	0	0	1.0	0
13–15	0.8	1.58	8.5	−1.4	3.64	−2.7	−1.16
14–15	−4.8	−9.45	5.1	8.6	21.84	2.34	7.24
15–16	8.3	16.34	−19.2	−14.7	37.77	−18.58	−12.28
16–17	−9.6	−18.90	16.1	16.9	−43.68	12.94	14.06
17–18	19.3	38.00	−32.6	−34.3	87.82	−26.34	−28.72
18–19	−15.2	−29.93	22.8	27.1	−69.17	16.72	22.74
19–20	33.1	65.17	−49.6	−59	150.62	−36.34	−49.5
20–21	−21.5	−42.33	30.2	38.4	−97.83	20.78	32.26
21–22	40.6	79.94	−57	−72.3	184.75	−39.2	−60.2

Notation: DL = dead load, *IL* = imposed load, *IP* = internal pressure, (−) = compression

The maximum values from the above load combinations are tabulated in Tables 11.4 and 11.6. These will be used later in the design of members.

Note that design conditions arising from notional horizontal loads specified in Clause 2.4.2.3 of BS 5950: Part 1 are not as severe as those in cases 2–5 in Table 11.4. The displacements at every joint are computed in the analyses, but only the critical values are of interest. They are summarized in Table 11.7 and compared with the maximum allowable values.

Table 11.5 Summary of member forces (kN) and moments (kN m) for columns and truss member connected to them

Member		Dead load (DL)	Imposed load (IL)	Crane load (CRWL)	Crane load (CRSL)	Wind at 0° (IS)	Wind at 0° (IP)	Wind at 90° (IP)
1–2(M)		−26.4	−51.9	−16.9	0.4	9.4	40.9	42.7
	Top	−22.1	−43.5	−7.1	−4	6.5	−24.1	−47.7
	Bottom	18.6	36.6	11.3	8.6	−35.1	8.3	54.1
2–3(M)		−26.4	−51.9	−0.7	0.4	8.8	38.5	40.7
	Top	−33.1	−64.9	4.2	4	−3.5	−38.9	−58.4
	Bottom	−22.1	−43.5	0.9	−4	6.5	−24.1	−47.7
3–4(M)		−26.4	−51.9	−0.7	−0.4	8.8	38.5	40.7
	Top	0	0	0	0	0	0	4
	Bottom	−33.1	−64.9	4.2	4	3.5	−38.9	−58.4
3–5		−40.5	−79.7	7.5	4	13.2	50.3	63
4–5		40.6	79.9	−1.7	−1	−14.5	−66.4	−72.3
4–6		−1.3	−2.6	−2.7	−3.1	−8.8	17.5	3
20–22		−1.3	−2.6	9.1	3.1	−7.2	17.5	3
21–22		40.6	79.9	2.4	1	−10.1	−57	−72.3
21–23		−40.5	−79.7	−7.8	−4	25.5	−73.4	−63
22–23(M)		−26.4	−51.9	−0.7	−0.4	5	30.8	40.7
	Top	0.0	0	0	0	0	0	0
	Bottom	−33.1	−64.9	1.1	4	−17.9	−61.9	−58.4
23–24(M)		−26.4	−51.9	−0.7	−0.4	5	30.8	40.7
	Top	−33.1	−64.9	1.1	4	−17.9	−61.9	−58.4
	Bottom	−22.1	−43.5	16	0.4	6.9	−42.7	−47.2
24–25(M)		−26.4	−51.9	−72.2	−0.4	5.2	32.1	42.7
	Top	−412.1	−43.5	−15.9	−4	6.9	−42.7	−47.7
	Bottom	18.6	36.6	2.4	8.6	−17.4	63.6	54.2

Table 11.6 Load combination for Table 11.5

	1.4 DL + 1.6 DL	1.4 DL + 1.6 LL + 1.6 CRWL	1.4 DL + 1.6 LL + 1.4 TCL	DL + 1.4x Wind at 0° (IS)	DL + 1.4x Wind at 0° (IP)	DL + 1.4 Wind at 90° (IP)	1.2(DL + TCL + Wind at 0° (IS)
	−120.00	−147.04	−146.4	−13.24	30.86	33.38	−40.20
Top	−100.54	−111.90	−118.3	−13.00	−55.84	−88.88	−32.04
Bottom	84.60	102.68	116.68	−30.54	30.22	94.34	4.08
	−120.00	−121.00	−120.48	−14.08	27.50	30.58	−21.48
Top	−150.18	−143.46	−137.1	−38.00	−87.56	−114.86	−34.08
Bottom	−100.54	−99.00	−105.5	−13.00	−55.84	−88.88	−22.44
	−120.00	−121.00	−121.76	−14.08	27.50	30.58	−22.44
Bottom	−150.18	−143.46	−137.06	−38.00	−87.56	−114.86	−34.08
	−184.22	−172.22	−165.82	−22.02	29.92	47.70	−18.96
	184.68	182.00	−180.36	20.30	−52.36	−60.62	28.08
	−5.98	−10.30	−15.26	−13.62	23.20	2.90	−19.08
	−5.98	8.58	13.54	−11.38	23.20	2.90	4.44
	184.68	188.52	190.12	26.46	−39.20	−60.62	40.68
	−184.22	−196.70	−197.34	−4.80	−143.26	−128.70	−27.84
	−120.00	−121.12	−121.76	−19.40	16.72	30.58	−27.00
Top	0	0	0	0	0.00	0.00	0.00
Bottom	−150.18	−132.58	−126.18	−58.16	−119.76	−114.86	−43.20
	−120.00	−121.10	−121.76	−19.40	30.58	−27.00	

(contd. overleaf)

Table 11.6 (contd.)

	1.4 DL + 1.6 DL	1.4 DL + 1.6 LL + 1.6 CRWL	1.4 DL + 1.6 LL + 1.4 TCL	DL + 1.4x Wind at 0° (IS)	DL + 1.4x Wind at 0° (IP)	DL + 1.4 Wind at 90° (IP)	1.2(DL + TCL + Wind at 0° (IS)
Top	−150.18	−132.58	−126.18	−56.16	119.76	−114.86	−43.20
Bottom	−100.54	−74.94	−68.54	−12.44	−81.88	−88.18	5.76
	−120.00	−235.52	−236.16	−19.12	18.54	33.38	−112.56
Top	−100.54	−125.98	−132.38	−12.44@	−81.88	−88.88	−42.12
Bottom	84.60	88.44	102.2	−5.76	107.64	94.48	14.64

Read member number from Table 11.5.

Table 11.7 Critical joint displacements

Load case: 1.0 DL + 1.0 LL	
Max. vertical deflection at joint 13	= 32.1 mm
Max. horizontal deflection at joints 2 and 24	= 3.7 mm
Horizontal deflection at joint 22	= 1.8 mm
Load case: 1.0 CRWL + 1.0 CRSL	
Max. horizontal deflection at joint 22	= 2.62 mm
Load case: 1.0 CRWL = 1.0 Wind at 0° I.S.	
Max. horizontal deflection at joint 22	= 7.86 mm

$$\text{Allowable vertical deflection} = L/200 = 20 \times 1000/200$$
$$= 100\,\text{mm} > 32.1 \text{ (satisfactory)}.$$
$$\text{Allowable horizontal deflection} = L/300 - 8 \times 1000/300$$
$$= 27\,\text{mm}$$
$$\text{Max. horizontal deflection} = 1.8 + 7.86$$
$$= 9.66\,\text{mm} < 16 \text{ (satisfactory)}.$$

11.5 Structural design of members

11.5.1 Design of the truss members

Using Grade S275 steel with a design strength of 275 N/mm^2, the truss members are designed using structural steel angle sections.

(1) Top chord members 10–12, 12–13, 13–14 and 14–16, etc.

Maximum compression from loads combination (2) = −248 kN,
Maximum tension from loads combination (5) = 82.6 kN.

Try 100 × 100 × 12.0 mm angle

$$r_\text{v} = 1.94\,\text{cm}, \quad A = 22.7\,\text{cm}^2.$$

The section is plastic. Lateral restraint is provided by the purlins and web members at the node points.

Slenderness $L/r_\text{v} = 2110/19.4 = 108$,

$p_\text{c} = 113\,\text{N/mm}^2$ (Table 24(c)).

Compression resistance:

$$p_c = 113 \times 22.7/10 = 256\,\text{kN} > 248\,\text{kN (satisfactory)}.$$

The section will also be satisfactory in tension.

(2) Bottom chord members 9–11, 11 –11 a, 11 a–15 and 15–17 etc.

Maximum tension from loads combination (1) $= 211.1\,\text{kN}$,
Maximum compression from loads combination (5) $= -82.3\,\text{kN}$.

Try $90 \times 90 \times 8.0\,\text{mm}$ angle
$$r_v = 1.76\,\text{cm}, \quad A = 13.9\,\text{cm}^2.$$

Tension capacity $= 275 \times 13.9/10 = 382\,\text{kN} > 211.1\,\text{kN}$.
Lateral support for the bottom chord are shown in Figure 11.1. The slenderness values are:

$$L_E/r_v = 2110/17.6 = 119,$$
$$L_E/r_x = 4200/27.4 = 153,$$
$$p_c = 66\,\text{N/mm}^2 \text{ (Table 24(c))},$$
Compression resistance $= 66 \times 13.9/10 = 92\,\text{kN} > 82\,\text{kN (satisfactory)}.$

All web members

Maximum tension $= 184.8\,\text{kN}$ (members 4–5),
Maximum compression $= -97.8\,\text{kN}$ (members 5–6).

Try $70 \times 70 \times 6.0\,\text{mm}$ angles

$$A = 8.13\,\text{cm}^2,$$
$$r_v = 1.37\,\text{cm}, \quad r_x = 2.13\,\text{cm}.$$

The slenderness is the maximum of:

$$L_E/r_v = 0.85 \times 1085/13.7 = 67.3,$$
$$L_E/r_x = (0.7 + 1085/21.3) + 30 = 65.6,$$
$$p_c = 186.4\,\text{N/mm}^2 \text{ (Table 24(c))},$$
$$p_c = 186.4 \times 8.13/10 = 151.5\,\text{kN} > 97.8\,\text{kN (satisfactory)}.$$

The angle is connected through one leg to a gusset by welding.

Net area $= 7.77\,\text{cm}^2$,
Tension capacity $= 275 \times 7.77/10 = 214\,\text{kN} > 184.8\,\text{kN (satisfactory)}.$

11.5.2 Column design

The worst loading condition for the column is that due to ultimate loads from dead load, imposed load and maximum crane load on members 24–25. The design loads are extracted from Table 11.6 as follows:

Maximum column compressive load $= -236$ kN,

Maximum moment at top of column $= -126$ kN m,

Corresponding moment at bottom $= 88$ kN m.

Try 457×191 UB67, the properties of which are:

$$A = 85.4 \, \text{cm}^3, \quad r_x = 18.55 \, \text{cm},$$
$$u = 0.873 \qquad r_y = 4.12 \, \text{cm},$$
$$x = 37.9 \qquad S_x = 1470 \, \text{cm}^3,$$
$$Z_x = 1296 \, \text{cm}^3.$$

Design strength $p_y = 275 \, \text{N/mm}^2$ (Table 9 of code)

Effective lengths $L_x = 1.5 \times 7000 = 10\,500$ mm (Appendix D of code),

$\qquad\qquad\qquad L_y = 0.85 \times 5500 = 4675$ mm (Appendix D of code).

Maximum slenderness ratio:

$\lambda = L_y/r_y = 4675/41.2 = 113.4$,

From Table 24(c) the value of $p_c = 105.6 \, \text{N/mm}^2$,

Value of $M_{cx} = S_x \times p_y = 1470 \times 275/1000 = 404$ kN m,

Check for $1.2 \times 275 \times 1296/1000 = 427$ kN m > 404.

$\beta = M_2/M_1 = -88/126 = -0.698$.

From Table 18, equivalent moment factor $m = 0.43$.

Equivalent moment $M' = m \times M_1 = 0.43 \times 126 = 54.2$ kN m.

Determine the value of buckling resistance moment M_b:

$\lambda_{LT} = uv\lambda$,

$\quad N = 0.5$,

with $\lambda/x = 11\,3.4/37.9 = 2.99$; from Table 19, $v = 0.91$.

$\lambda_{LT} = 0.873 \times 0.91 \times 113.4 = 90.1$,

From Table 16, the value of $p_b = 143.8 \, \text{N/mm}^2$,

$M_b = S_x \times p_b = 1470 \times 143.8/1000 = 211.4$ kN m.

Check for local capacity at top of column:

$$\frac{F}{A_g \times p_y} + \frac{M_x}{M_{cx}} < 1$$

i.e. $\dfrac{236 \times 10}{85.4 \times 275} + \dfrac{126}{404} = 0.1 + 0.31 = 0.41 < 1$ (satisfactory)

Similarly, check for local capacity at bottom of column. The interaction criteria is equal to 0.317, which is satisfactory. Check the stanchion for overall buckling:

$$\frac{F}{A_g \times p_c} + \frac{M}{M_b} = \frac{0.26 + 0.26}{85.4 \times 105.6} = \frac{54.2}{211.4} = 0.52 < 1 \text{ (satisfactory)}$$

The 457×191 UB67 provided is satisfactory. The reader may try with a 457×152 UB60 to increase the stress ratio and achieve greater economy. For the design of the crane girder, the reader should refer to Chapter 4.

11.6 Steelwork detailing

The details for the main frame and connections are presented in Figure 11.9.

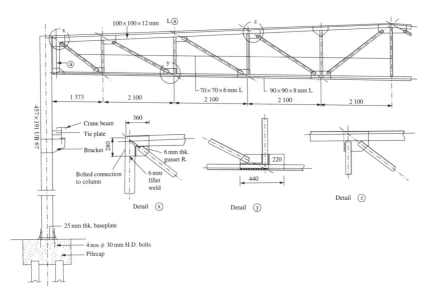

Figure 11.9 Steel frame details

Steelwork detailing

12.1 Drawings

Drawings are the means by which the requirements of architects and engineers are communicated to the fabricators and erectors, and must be presented in an acceptable way. Detailing is given for selected structural elements.

Drawings are needed to show the layout and to describe and specify the requirements of a building. They show the location, general arrangement and details for fabrication and erection. They are also used for estimating quantities and cost and for making material lists for ordering materials.

Sufficient information must be given on the designer's sketches for the draughtsman to make up the arrangement and detail drawings. A classification of drawings is set out below:

Site or location plans. These show the location of the building in relation to other buildings, site boundaries, streets, roads, etc.

General arrangement. This consists of plans, elevations and sections to set out the function of the building. These show locations and leading dimensions for offices, rooms, work areas, machinery, cranes, doors, services, etc. Materials and finishes are specified.

Marking plans. These are the framing plans for the steel-frame building showing the location and mark numbers for all steel members in the roof, floors and various elevations.

Foundation plans. These show the setting out for the column bases and holding-down bolts and should be read in conjunction with detail drawings of the foundations.

Sheeting plans. These show the arrangement of sheeting and cladding on building.

Key plan. If the work is set out on various drawings, a key plan may be provided to show the portion of work covered by the particular drawing.

Detail drawings. These show the details of structural members and give all information regarding materials, sizes, welding, drilling, etc. for fabrication. The mark number of the detail refers to the number on the marking plan.

Detail drawings and marking plans will be dealt with here.

12.2 General recommendations

12.2.1 Scales, drawing sizes and title blocks

The following scales are recommended:

Site, location, key plans 1:500/1:200,
General arrangement 1:200/1:100/1:50,
Marking plans 1:200/1:100,
Detail drawings 1:25/1:10/1:5,
Enlarged details 1:5/1:2/1:1.

The following drawing sizes are used:

A4 210 × 297—sketches,
A3 297 × 420—details,
A2 420 × 59—general arrangement, details,
A1 594 × 841—general arrangement, details,
A0 841 × 1189—general arrangement, details.

Title blocks on drawings vary to suit the requirements of individual firms and authorities. Typical title blocks are shown in Figure 12.1.

Materials lists can either be shown on the drawing or on separate A4 size sheets. These generally give the following information:

Item or mark number,
Description,
Material,
Number off,
Weight,
Etc.

12.2.2 Lines, sections, dimensions and lettering

Recommendations regarding lines, sections and dimensions are shown in Figure 12.2.

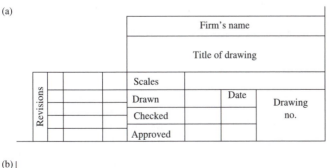

Figure 12.1 Typical title blocks

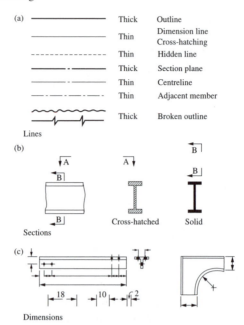

Figure 12.2 Recommendations for lines, sections and dimensions

Section		Reference	Example
Universal beam	I	UB	610×178×91 kg/m UB
Universal Column	I	UC	203×203×89 kg/m UC
Joist	I	RSJ Joist	203×102×25.33 kg/m Joist
Channel	[	channel	254×76×28.29 kg/m [
Angle	L	Angle	150×75×10 L
Tee	T	Tee	178×203×37 kg/m struct. Tee
Rectangular hollow section	☐	RHS	150.4×100×6.3 RHS
Circular hollow section	○	CHS	76.1 O.D.×5 CHS

Figure 12.3 Representation of rolled and formed steel sections

12.3 Steel sections

Rolled and formed steel sections are represented on steelwork drawings as set out in Figure 12.3. The first two figures indicate the size of section, (for example, depth and breadth). The last figure indicates the weight in kg/m for beams, columns, channels and tees. For angles and hollow sections, the last figure gives the thickness of steel. With channels or angles, the name may be written or the section symbols used as shown. Built-up sections can be shown either by:

(1) true section for large scale views, or
(2) diagrammatically by heavy lines with the separate plates and sections separated for clarity for small-scale views. Here, only the depth and breadth of the section may be true to scale.

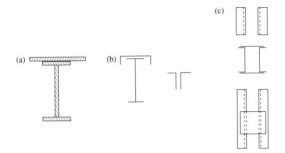

Figure 12.4 Representation of built-up sections

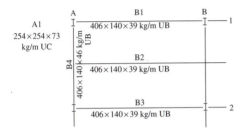

Figure 12.5 Representation of beams and columns

These two cases are shown in Figure 12.4. The section is often shown in the middle of a member inside the break lines in the length, as shown in (c). This saves having to draw a separate section.

Beams may be represented by lines, and columns by small-scale sections in heavy lines, as shown in Figure 12.5. The mark numbers and sizes are written on the respective members. This system is used for marking plans.

12.4 Grids and marking plans

Marking plans for single-storey buildings present no difficulty. Members are marked in sequence as follows:

Columns AI, A2, ... (See grid referencing below)
Trusses T I, T2, ...
Crane girders CG 1, CG2, ...
Purlins PI, P2, ...
Sheeting rails SRI, SR2, ...
Bracing Bl, B2, ...
Gable columns GS 1, GS2, ...
etc.

Various numbering systems are used to locate beams and columns in multi-storey buildings. Two schemes are outlined below:

(1) In plan, the column grid is marked A, B, C, ... in one direction and 1, 2, 3, ... in the direction at right angles. Columns are located A1, C2, ... Floors are numbered A, B, C, ... for ground, first, second, ..., respectively.

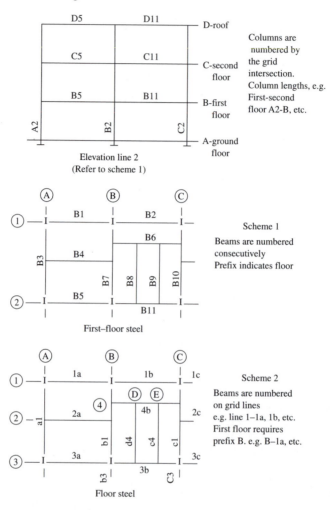

Figure 12.6 Marking Systems for multi-storey buildings

Floor beams (for example, on the second floor) are numbered B1, B2, . . .
Column lengths are identified: for example, A4-B is the column on grid
intersection A4, length between second and third floors.
(2) A grid line is required for each beam.

The columns are numbered by grid intersections as above.
The beams are numbered on the grid lines with a prefix letter to give the
floor if required. For example,

Second floor—grid line 1—C-1a, C-1b,—
Second floor—grid line B—C-1b, C-b2,—

The systems are shown in Figure 12.6. The section size may be written on the
marking plan.

12.5 Bolts

12.5.1 Specification

The types of bolts used in steel construction are:

Ordinary Grade 4.6 or black bolts,
High strength Grade 8.8 bolts,
Preloaded HSFG friction-grip bolts.

The British standards covering these bolts are:

BS 4190: ISO Metric Black Hexagon Bolts, Screws and Nuts,
BS 4395: Part 1: High Strength Friction-Grip Bolts: General Grade,
BS 41604: Part 1: The Use of High Strength Friction-Grip Bolts in Structural
    Steelwork: General Grade.

The strength grade designation should be specified. BS 5950 gives the
strengths for ordinary bolts for grades 4.6 and 8.8. Minimum shank tensions
for friction-grip bolts are given in BS4604. The nominal diameter is given in
millimetres. Bolts are designated as M12, M16, M20, M229, M24, M27, M30,
etc., where 12, 16, etc. is the diameter in millimetres. The length under the head
in millimetres should also be given.

Examples in specifying bolts are as follows:

4 No. 16 mm dia. (or M16) black hex. hd. (hexagon head) bolts, strength
    grade,
4.6 × 40 mm length,
20 No. 24 mm dia. friction-grip bolts × 75 mm length.

The friction-grip bolts may be abbreviated HSFG. The majority of bolts may
be covered by a blanket note. For example:

All bolts M20 black hex. hd.
All bolts 24-mm dia. HSFG unless otherwise noted.

12.5.2 Drilling

The following tolerances for drilling are used:

Ordinary bolts-holes to have a maximum of 2 mm clearance for bolt dia-
    meters up to 24 mm and 3 mm for bolts of 24 mm diameter and over.
For friction-grip bolts, holes are drilled the same as set out above for ordinary
    bolts. Hole diameters are given in Table 36 of BS 5950: Part 1.

Drilling may be specified on the drawing by notes as follows:

All holes drilled 22 mm dia. for 20 mm dia. ordinary bolts.
All holes drilled 26 mm dia., unless otherwise noted.

12.5.3 Designating and dimensioning

The representation for bolts and holes in plan and elevation on steelwork
drawings is shown in Figure 12.7(a). Some firms adopt different symbols
for showing different types of bolts and to differentiate between shop and
field bolts. If this system is used, a key to the symbols must be given on the
drawing.

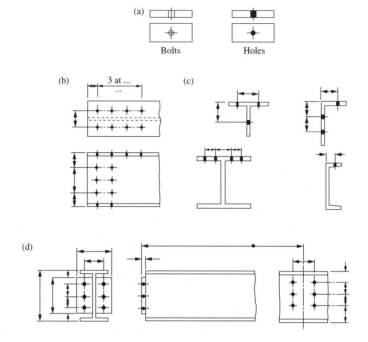

Figure 12.7 Representation of bolts and holes on steelwork drawings

Gauge lines for drilling for rolled sections are given in the *Structural Steelwork Handbook*. Dimensions are given for various sizes of section, as shown in Figure 12.7(c). Minimum edge and end distances were discussed in Section 4.2.2 above.

Details must show all dimensions for drilling, as shown in Figures 12.7(b) and (d). The holes must be dimensioned off a finished edge of a plate or the back or end of a member. Holes are placed equally about centre lines. Sufficient end views and sections as well as plans and elevations of the member or joint must be given to show the location of all holes, gussets and plates.

12.6 Welds

As set out in Chapter 4, the two types of weld are butt and fillet.

12.6.1 Butt welds

The types of butt weld are shown in Figure 12.8 with the plate edge preparation and the fit-up for making the weld. The following terms are defined:

T = thickness of plate,
g = gap between the plates,
R = root face,
a = minimum angle.

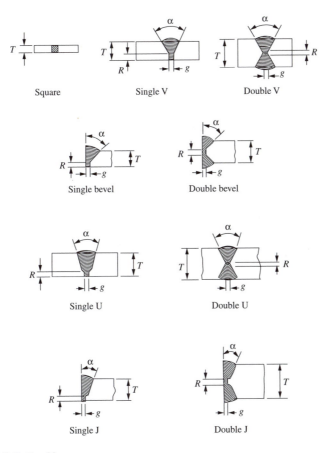

Figure 12.8 Butt welds

Values of the gap and root face vary with the plate thickness, but are of the order of 1–4 mm. The minimum angle between prepared faces is generally 50–60° for V preparation and 30–40° for U preparation. For thicker plates, the U preparation gives a considerable saving in the amount of weld metal required. The student should consult the complete details given in BS 499: 1965: Part 2, *Welding Terms and Symbols*.

Welds may be indicated on drawings by symbols from Table 1 of BS 499 and these are shown in Figure 12.9. Using these symbols, butt welds are indicated on a drawing as shown in Figure 12.10(a). Reference should be made to BS 499 for a complete set of examples using these symbols.

The weld name may be abbreviated: for example, DVBW for double V butt weld. An example of this is shown in Figure 12.8(b). Finally, the weld may be listed by its full description and an enlarged detail given to show the edge preparation and fit-up for the plates. This method is shown in Figure 12.8(c). Enlarged details should be given in cases where complicated welding is required. Here, detailed instructions from a welding engineer may be required and these should be noted on the drawing.

Fillet	◿
Square butt weld	╥
Single V butt weld	▽
Double V butt weld	⧓
Single U butt weld	∪
Double U butt weld	ੲ
Single bevel weld	ↁ
Double bevel weld	Ҟ
Single J weld	↿
Double J weld	↻

Figure 12.9 Symbols for welds

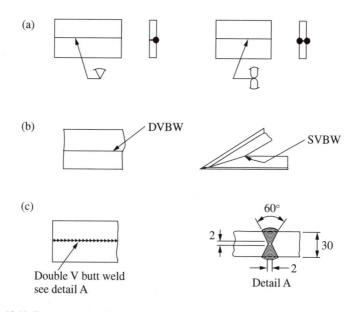

(a)

(b)   DVBW   SVBW

(c)   Double V butt weld
       see detail A

       60°
       2
       30
       2
       Detail A

Figure 12.10 Representation of butt welds

12.6.2 Fillet welds

These welds are triangular in shape. As set out in Chapter 10, the size of the weld is specified by the leg length. Welds may be indicated symbolically, as shown in Figure 12.11(a). BS 499 should be consulted for further examples. The weld size and type may be written out in full or the words 'fillet weld' abbreviated, (for example, 6 mm FW). If the weld is of limited length, its exact location should be shown and dimensioned. Intermittent weld can be shown by writing the weld size, then two figures which indicate length and space between welds. These methods are shown in Figure 12.11(b). The welds can also be shown and specified by notes, as on the plan view in Figure 12.11(b). Finally, a common method of showing fillet welds is given in Figure 12.11(c), where thickened lines are used to show the weld.

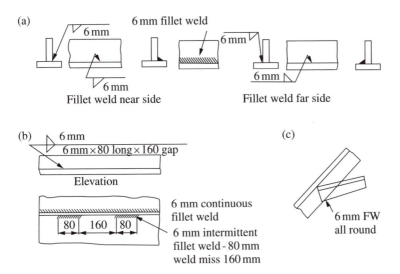

Figure 12.11 Representation of fillet welds

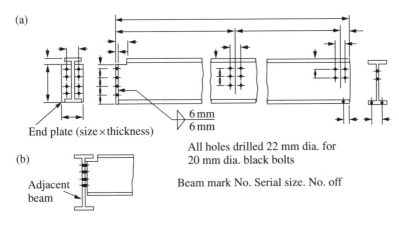

Figure 12.12 Beam details

12.7 Beams

Detailing of beams, purlins and sheeting rails is largely concerned with showing the length, end joints, welding and drilling required. A typical example is shown in Figure 12.12. Sometimes it is necessary to show the connecting member and this may be shown by chain dash lines, as shown in Figure 12.12(b).

12.8 Plate girders

The detailed drawing of a plate girder shows the girder dimensions, flange and web plate sizes, sizes of stiffeners and end plates, their location and the details for drilling and welding. Any special instructions regarding fabrication

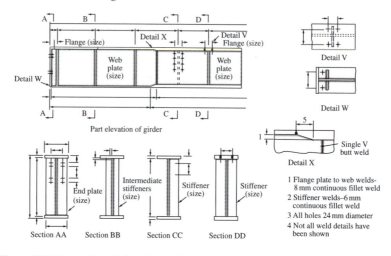

Figure 12.13 Typical details for plate girder

should be given on the drawing. For example, preheating may be required when welding thick plates or 2 mm may require machining off flame-cut edges to reduce the likelihood of failure by brittle fracture or fatigue on high yield-strength steels.

Generally all information may be shown on an elevation of the girder together with sufficient sections to show all types of end plates, intermediate and load-bearing stiffeners. The elevation would show the location of stiffeners, brackets and location of holes and the sections complete this information. Plan views on the top and bottom flange are used if there is a lot of drilling or other features best shown on such a view. The draughtsman decides whether such views are necessary.

Part sectional plans are often used to show stiffeners in plan view. Enlarged details are frequently made to give plate weld edge preparation for flange and web plates, splices and for load-bearing stiffeners that require full-strength welds.

Notes are added to cover drilling, welding and special fabrication procedures, as stated above. Finally, the drawing may contain a material list giving all plate sizes required in the girder. Typical details for a plate girder are shown in Figure 12.13.

12.9 Columns and bases

A typical detail of a column for a multi-storey building is shown in Figure 12.14. The bottom column length with base slab, drilling for floor beams and splice details is shown.

A compound crane column for a single-storey industrial building is shown in Figure 12.15. The crane column is a built-up section and the roof portion is a universal beam. Details at the column cap, crane girder level and base are shown.

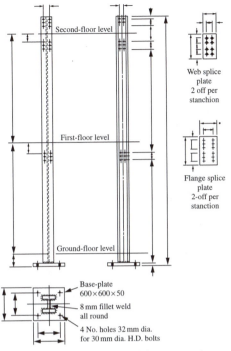

All holes 22 mm dia. for 20 mm dia. black bolts except as noted

Figure 12.14 Stanchion in a multi-storey building

12.10 Trusses and lattice girders

The rolled sections used in trusses are small in relation to the length of the members. Several methods are adopted to show the details at the joints. These are:

(1) If the truss can be drawn to a scale of 1 in 10, then all major details can be shown on the drawing of the truss.
(2) The truss is drawn to a small scale, 1 in 25, and then separate enlarged details are drawn for the joints to a scale of 1 in 10 or 1 in 5.

Members should be designated by size and length; for example, where all dimensions are in millimetres:

100 × 75 × 10 Angle × 2312 long,
100 × 50 × 4 RHS × 1310 long.

On sloping members, it is of assistance in fabrication to show the slope of the member from the vertical and horizontal by a small triangle adjacent to the member.

The centroidal axes of the members are used to set out the frame and the members should be arranged so that these axes are coincident at the nodes of the truss. If this is not the case, the eccentricity causes secondary stresses in the truss.

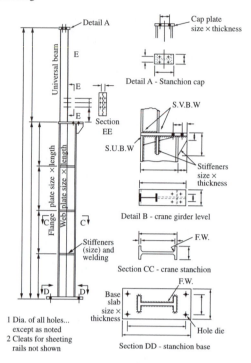

Figure 12.15 Compound crane stanchion for a single-storey industrial building

Full details are required for bolted or welded splices, end plates, column caps, etc. The positions of gauge lines for drilling holes are given in the *Structural Steelwork Handbook*. Dimensions should be given for edge distances, spacing and distance of holes from the adjacent node of the truss. Splice plates may be detailed separately.

Sometimes the individual members of a truss are itemized. In these cases, the separate members, splice plates, cap plates, etc. are given an item number. These numbers are used to identify the member or part on a material list.

The following figures are given to show typical truss detailing:

Figure 12.16 shows a portion of a flat roof truss with all the major details shown on the elevation of the truss. Each part is given an item number for listing.

Figure 12.17 shows a roof truss drawn to small scale where the truss dimensions, member sizes and lengths are shown. Enlarged details are given for some of the joints.

12.11 Computer-aided drafting

Computer-aided drafting (CAD) is now being introduced into civil and structural drafting practice. It has now replaced much manual drafting work. CAD can save considerable man-hours in drawing preparation, especially where standard details are used extensively.

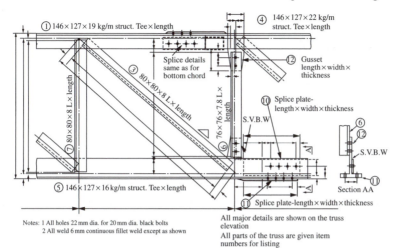

Notes: 1 All holes 22 mm dia. for 20 mm dia. black bolts
      2 All weld 6 mm continuous fillet weld except as shown

All major details are shown on the truss elevation
All parts of the truss are given item numbers for listing

Figure 12.16 Portion of a flat roof truss

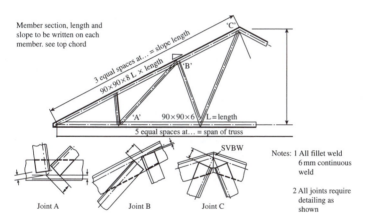

Figure 12.17 Small-span all-welded truss

In computerized graphical systems, the drawing is built up on the screen, which is divided into a grid. A menu gives commands (for example, line, circle, arc, text, dimensions, etc.). Data are input through the digitizer and keyboard. All drawings are stored in mass storage devices from which they can be retrieved for subsequent additions or alterations. Updating of drawings using the CAD system can be accomplished with little effort. Standard details used frequently can either be drawn from a program library or created and stored in the user's library.

Some comprehensive steel design and detailing software can automate the entire process for standard type of steel structures.

Overlay of drawings in CAD software is a very useful feature, which allow repeated usage of common drawing templates (for example, grid lines, floor plans, etc.) and results in considerable time saving by reducing input time required. Texts and dimensions can be typed from the keyboard using appropriate character sizes.

References

1 RICHARDS, K.G., *Fatigue Strength of Welded Structures*, The Welding Institute, Cambridge, 1969
2 BOYD, G.M., *Brittle Fracture in Steel Structures*, Butterworths, London, 1970
3 *Building Regulations and Associated Approved Documents B*, HMSO, London, 2000
4 *Fire Protection of Structural Steel in building* (Third Edition), ASFPCM, 2002
5 CASE, J. and CHILER, A.H., *Strength of Materials*, Edward Arnold, London, 1964
6 TRAHAIR, N.S., BRADFORD, M.A. and NETHERCOT, D.A., *The Behaviour and Design of Steel Structures to BS5950*, Spon Press, 2001
7 NETHERCOT, D.A. and LAWSON, R.M., *Lateral Stability of Steel Beams and Columns – Common Cases of Restraint*, Steel Construction Institute, 1992
8 HORNE, M.R., *Plastic Theory of Structures*, Pergamon Press, Oxford, 1979
9 *Steel Designers Manual*, Steel Construction Institute and Blackwell Science, 2003
10 GHALI, A. and NEVILLE, A.M., *Structural Analysis*, Chapman and Hall, London, 1989
11 *Joints in Steel Construction: Simple Connections*, SCI & BCSA, 2002
12 WARD BUILDINGS COMPONENTS, *Multibeam Products*, Ward Buildings Components Ltd, 2002
13 TIMOSHENKO, S.P. and GERE, J.M., *Theory of Elastic Stability*, McGraw-Hill, New York, 1961
14 PORTER, D.M., ROCKEY, K.C. and EVANS, H.R., *The Collapse Behaviour of Plate Girders Loaded in Shear*, The Structural Engineer, August 1975
15 PRATT, J.L., *Introduction to the Welding of Structural Steelwork*, Steel Construction Institute, 1989
16 HOLMES, M. and MARTIN, L.H., *Analysis and Design of Structural Connections—Reinforced Concrete and Steel*, Ellis Horwood, London, 1983
17 SALMON, C.G. *et al.*, *Laboratory Investigation on Unstiffened Triangular Bracket Plates*, ASCE Structural Division, April 1964
18 HORNE, M.R. and MORRIS, L.J., *Plastic Design of Low Rise Frames*, Collins, London, 1981
19 COATES, R.C., COUTIE, M.G. and KONG, F.K., *Structural Analysis*, Spon E & FN, 1998
20 SOUTHCOMBE, C., *Design of Structural Steelwork: Lattice Framed Industrial Buildings*, Steel Construction Institute, 1993

Index